COMBINED PRODUCTION OF HEAT AND POWER (COGENERATION)

Proceedings of a European seminar organised by

— the Commission of the European Communities, Directorate-General for Energy

and

— the Instituto para la Diversificacion y Ahorro de la Energia (IDAE)

with the cooperation of

— Gomez Pardo Foundation's Energy Commission

and held in Madrid, Spain, 10–11 October 1989.

Particular thanks are due to Mr L. Arimany de Pablos (IDAE), consultant to the Commission of the European Communities, for editorial assistance.

COMBINED PRODUCTION OF HEAT AND POWER (COGENERATION)

Edited by

J. SIRCHIS

Directorate-General for Energy, Commission of the European Communities, Brussels, Belgium

Routledge
Taylor & Francis Group
LONDON AND NEW YORK

First published 1990 by Elsevier Science Publishers Ltd.

Published 2020 by Routledge
2 Park Square, Milton Park, Abingdon, Oxon OX14 4RN
52 Vanderbilt Avenue, New York, NY 10017

First issued in paperback 2020

Routledge is an imprint of the Taylor & Francis Group, an informa business

Sole Distributor in the USA and Canada
ELSEVIER SCIENCE PUBLISHING CO., INC.

WITH 36 TABLES AND 51 ILLUSTRATIONS

British Library Cataloguing in Publication Data

Combined production of heat and power (cogeneration).
1. Combined heat and power-schemes
I. Sirchis, J.
333.793

Library of Congress CIP data applied for

Publication arrangements by Commission of the European Communities, Directorate-General Telecommunications, Information Industries and Innovation, Scientific and Technical Communication Unit, Luxembourg.

EUR 12714

ISBN 13: 978-0-367-58013-1 (pbk)
ISBN 13: 978-1-85166-524-2 (hbk)

PREFACE

The existence of significant uncertainty as to the long term prospects for energy supply and demand, following the rapid fall in oil prices, has stimulated both the international energy situation as well as that of the Community and made it essential that the substantial progress already made in restructuring the Community's energy economy be maintained and, if necessary, reinforced.

The European Energy Policy objectives for the year 1995 call for adequate energy supply, controlled energy prices and increased environmental concern. All these constraints necessitate the rational exploitation of the primary energy forms by the EEC member states.

The above objectives can be attained either by energy saving or increased energy efficiency, or finally through the development of new technologies to augment both saving and efficiency. Better insulation heat and material recycling, or application of improved processes, are typical examples.

Cogeneration is the process of generating electricity with synchronous utilisation of the useful heat wastes produced. Thus the overall efficiency is increased, the consumption of primary energy is lowered and the sequential pollutant emissions are eliminated.

Much progress has been made up to now in the field of cogeneration in the EEC countries in industry, power plants, district heating and building air-conditioning, but the necessary expansion of applications requires further technological progress, new methods of financing and an appropriate legislative basis of reference.

CONTENTS

Preface v

OPENING SESSION

Opening Address
PEREZ PRIM, General Director for Energy, Spanish Ministry of Industry and Energy, Madrid, Spain 2

Welcoming Address
F. SERRANO, General Director, I.D.A.E., Madrid, Spain 4

Introductory Speech: 'Energy Policy of the Commission of the European Communities'
F. KINDERMANN, Head of Division, Commission of the European Communities, Directorate-General for Energy, Energy Technology Directorate, Programme Management: Solid Fuel and Energy Saving 5

OVERVIEW OF TECHNOLOGIES

Cogeneration Technologies: Present and Future Developments
F. ALBISU, Sener, Igenieria y Sistemas S.A., Bilbao-Madrid, Spain 10

Cogeneration and Wood/Biomass Fueled Power Systems
E.P. GYFTOPOULOS, Massachusetts Institute of Technology, Cambridge, Massachusetts, USA 20

State of Co-generation in Spain
D. CONTRERAS, A. GOMEZ-ANGULO, Department for Cogeneration and Substitution, I.D.A.E., Madrid, Spain 35

Summary of Discussions 59

COGENERATION FINANCING AND LEGISLATION IN E.E.C. AND THIRD COUNTRIES

The History and Status of Financing Cogeneration Projects in California with Prospects for the Future
J. HAMRIN, Independent Energy Producers Association, Jan Hamrin Associates, Mill Valley, California, USA 62

Third Party Financing
D.A. FEE, Directorate-General for Energy, Commission of the European Communities 74

Comparative Analysis of the Legal Conditions in the Non-EEC Industrialised Countries: Difficulties and Advantages
D. DRISCOLL, Faculty of Law, University College, Galway, Ireland 85

Summary of Discussions 94

ROUND TABLE ON COGENERATION AND ENVIRONMENT

J. SIRCHIS, Directorate-General for Energy, Commission of the European Communities; A. DIAZ VARGAS, Directorate-General for Environment, Ministry of Public Works, Spain; D. DRISCOLL, I.E.A.; D.A. FEE, Commission of the European Communities; D. GREEN, Combined Heat and Power Association, UK; E. GYFTOPOULOS, M.I.T., USA; J.G. HAMRIN, Independent Energy Producers Association, USA 98

Summary of Discussions 108

COGENERATION IN EUROPEAN COMMUNITIES' MEMBER STATES

The Experience of One Enterprise
J.J. CAPARROS, Papelera del Jarama, S.A., Madrid, Spain 112

The Cogenerative Diesel Brescia Nord Afterburning Experience
G. MARANIELLO, Aerimpianti-Ansaldo, Milano, Italy 117

Midlife Conversion of a Waste Combustion Plant at Duiven, The Netherlands
F.W. BERKELMANS, P.G. KLOP, F.J. TERMÖHLEN, The Netherlands 127

Technical and Economic Aspects of CHP at Pfizer
P.P. McGLADE, Pfizer Chemical Corporation, Cork, Ireland 139

Hundred Thousand Hours Baseload Cogeneration with the IM-5000
E. HOLLROTTER, Dow Stade GmbH, Germany 151

Hundested Decentralized Heat and Power Plant
P. LOETH, Elselskabet EFFO, Denmark 164

Central 9.34 MW Electricity, Heating and Cooling Cogeneration Plant
C. FOUNAND COLL, L. MONTALT ROS, Nurel S.A., Spain 181

Summary of Discussions 190

Conclusions 193

LIST OF PARTICIPANTS 195

INDEX OF AUTHORS 211

OPENING SESSION

OPENING ADDRESS

Perez Prim
General Director for Energy,
Spanish Ministry of Industry and Energy

Ladies and Gentlemen.

Good morning and welcome to you all.

Before going any further, I must apologize right off for the absence the Secretary General for Energy and Mineral Resources and President of IDAE owing, not to a lack of interest concerning the subject of this Seminar, but to the physical impossibility of being in two places at the same time.

As Director General for Energy, it gives me great satisfaction to be able to open this European Seminar on Co-generation, which is the result of the joint work undertaken by the Institute for the Diversification and Saving of Energy, IDAE, and the Comission of the European Community together with the invaluable help and co-operation received from the Gómez Pardo Foundation.

I feel that this is a good time to review the progress made up to now in the field of co-generation in the EEC countries, among which Spain has not lagged behind; and this can clearly be seen from the fact that, where as the 1986 rate of growth in the use of this system was 8 % over the previous year, in 1987 this had gone up to 13 % and has continued to rise ever since. They will to continue in this direction is all too evident.

Primary energy saving derived from the joint generation of steam and electricity is of the greatest importance as regards national energy policy, in that it enables electricity to be produced with high rates of yield, since, for each electric KWh produced it burns on average only 50 % of the fuel which would otherwise be used in a conventional thermal power station.

However, not only does co-generation provide these advantages of energy saving at a national level, it also brings profits directly accruing to the company plus an increase in competitive edge which gives the company concerned the chance of winning a greater market share.

An aspect worth mentioning in the case of Spain, is the participation of electricity companies in the development of co-generation schemes; this will doubtless provide a degree of diversification of business and flexibility in consumer relations, which will be felt in the form of synergy that will in turn lend impetus to the efficiency of the plants in question.

It is on account of the obvious advantages to be gained from the steady advance of co-generation, that the Spanish Government, in accordance with the directives of the European Economic Commission, is giving a new boost to this kind of installation and is putting the finishing touches to a new body of law, soon to be passed, which will complement and fine-tune the 1982 legislation. We trust that, with the aid of the new regulations, we shall be able to attain the targets that have been set in the promising program drawn up for the future.

I hope, and have no doubts as to it being otherwise, that the topics raised during the Seminar will go to help the exchange of ideas between manufacturers and potential end-users and will be of special interest to managers with experience in energy matters as well as to researchers and students.

It only remains for me to say once again that I wish you every success, that I hope the sessions prove profitable and that, especially in the case of visitors from abroad, you all enjoy your stay in Madrid.

Thank you for your kind attention. I now formally declare the European Seminar on Co-generation to be open.

WELCOMING ADDRESS

F. SERRANO
General Director, IDAE

In my capacity as Director General of IDAE I should like to thank you for attending this Seminar and wish you a profitable exchange of ideas.

I should like to point out that, after the success of the 1st. International Co-generation Congress held in Madrid in 1988 and in view of IDAE's willingness to hold a second meeting of a similar nature in 1990, this Seminar serves to span both occasions; and, in doing so, it affords us the dual opportunity of discussing the present panorama of co-generation in European industry and looking at both the achievements recorded to date as well as the future, awaiting a technology, which, doubtless owing to the advantages it provides, is experiencing a boom.

It must be said that, within the context of its scope of activities, which are fundamentally aimed at promoting the efficient use of energy, IDAE has drawn up a specific program for the purposes of promoting co-generation technology in Spanish industry. In this respect, the results have been spectacular. Suffice to say that, between 1988 and 1989, 24 co-generating plants possessing a power capacity of 83 MW have been installed. Moreover, there are a further eighteen plants having a capacity of 126 MW which are now under construction and will be coming on stream within the next six months. On balance, this means that in the space of only two years, 42 installations with a power capacityy of 209 MW, will have been set up. With respect to 1987, this represents a 56 % rise in electric energy produced by co-generation, with a 65 % increase in the number of plants and 28 % increase in power capacity. Between 1988 and 1989, IDAE has played its part in this process of growth by participating directly in 13 schemes having a 33 MW power capacity, which has meant an investment of 3,000 million pesetas.

The goal of IDAE's program for the promotion of co-generation is to increase the plant power capacity thus installed by an additional 700 MW by the end of 92. The accumulated sum total of investment corresponding there will amount to aproximately 100,000 million pesetas and the electric energy produced by the new co-generating systems will mean a primary energy saving of 500,000 tep/per annum and a rise from the present figure of 2 % to that of 4 % in the level of electricity produced by means of co-geneation.

And on that note, I should just like to welcome you all once again. Thank you for your time and attention.

INTRODUCTORY SPEECH

"ENERGY POLICY OF THE COMMISSION OF THE EUROPEAN COMMUNITIES"

by

F. KINDERMANN, Head of Division
Commission of the European Communities
Directorate-General for Energy
Energy Technology Directorate
Programme Management: Solid Fuels and Energy Saving

If one goes back to the roots of the European Community, one discovers that two of the three Treaties deal, partly or completely, with energy.

- The Treaty establishing the EUROPEAN COAL AND STEEL COMMUNITY (ECSC) was signed in Paris in 1951.
- The Treaty establishing the EUROPEAN ATOMIC ENERGY COMMUNITY (EAEC or EURATOM) was signed in Rome in 1957.

Therefore, one could say that, from the beginning, the founders of Europe regarded energy as a very important brick for the construction of a real Community. In fact, one could say that most of the integrated Common Market has already been realised for coal, steel and uranium.

In spite of this, I must admit that there was virtually no general common energy policy existing until the first oil crisis back in 1973. It was only under the influence of this shock that quantified targets for selected energy carriers were defined. Of course, the main concern was, at that time, to substitute oil and to reduce the dependency of the Community. Therefore, solid fuels and energy efficiency played a very important role, and it should be noted that both provided the framework of the subsequent development of cogeneration, which is today's subject.

But let me come back to European Energy Policy. Once established, it led very quickly to tangible results. In fact, the consumption of imported oil was halved within 10 years, from 62% in 1973 to 31% in 1985. This forced the Commission to propose new targets for 1995, which were finally adopted by the Council in September 1986.

I will not go into these in great detail as we all know very well that, since then, conditions on the energy market have changed drastically: oil prices went down, as did coal prices on the world market; natural gas is pressing for a higher market share; and in some countries, nuclear energy continues to expand. In addition to this, there is more and more concern about the environment and particularly about the so-called greenhouse effect. You will certainly understand that all this gave reason to review the 1995 targets and will, most likely, lead the Commission to propose new targets for 2000 or 2005. As the outcome of this exercise is not yet predictable, I would like to mention today only three of the presently revised targets which may be of importance to cogeneration.

- Energy efficiency will remain one of the most important topics of Energy Policy, for the reasons of economy as well as of environment.

- Solutions are needed to establish a well-balanced relationship between Energy and the Environment. This will certainly become even more important in future and will require adequate solutions.

- Technology will have to play an extremely important role in achieving the targets.

It is quite interesting to see that these three items were amongst the Community's targets from the beginning. Yet, importance shifted from aspects of substitution and economics to the protection of the environment. We will have to see later how this may affect cogeneration but I feel obliged to say a few words first on the Integrated Market for Energy or, in short, 1992.

In fact, National as well as Community policies have to change to meet the situation that will exist after 1992. Energy is an area where this transition now has to be made in order to have the integrated European energy market followed by a true common energy policy at Community level.

The integration of Europe's internal energy market is now underway, and a number of new initiatives in this field have been launched since the beginning of 1989. These include new schemes for greater cross-frontier trade and competition in the gas and electricity sectors, a mechanism for taking into account the European dimension in the planning of major energy investments, and a new system allowing the transparency of gas and electricity prices. Other measures to ensure the 1992 deadline will follow.

In the longer term, it will be the Commission's task to propose to the Member States, a concise framework for an effective Community energy policy. Therefore, a new review of longterm energy prospects presently underway i.e., the 2010 study, it is too early to predict what the exact results of this study will be, but one can certainly expect that one of the major problems for the Community will be the impact on the environment of energy production and use. This means that all measures allowing a reduction of energy consumption will continue to have highest priority, and since cogeneration is among the most promising areas of energy conservation, it may be useful to briefly present to you what the Community has done so far.

More than ten year ago the Commission of the European Communities decided to submit a proposal to the Member States concerning the promotion of combined heat and power production and the recovery of waste heat.

The Council agreed to this initiative and adopted the recommendation[1)] that the Member States create advisory bodies or committees with the tasks of giving an opinion on all measures likely to lead to increased efficiency in the supply of heat for industry and promote the use of district heat supply systems. These committees were invited to consider specific measures such as, for example:

1) O.J. No.L 295, 18.11.1977, p. 5

- the identification and abolition of legal, administrative and price obstacles to the development of combined heat and power production;

- encouragement of combined heat and power production and heat transport schemes within the limits set by the EEC competition rules (Article 92 of the EEC Treaty);

- the provision of better information to small and medium sized industrial enterprises.

Furthermore, it was recommended that the Member States investigate and promote technical and economic studies and that they inform the Commission regularly of the measures taken in this field and of the results obtained or expected from these measures. Last - but not least - the advisory bodies should have regular exchanges of experience and should cooperate at Community level.

The Council recommendation led to quite a lot of activity in the following years and triggered the development of CHP and district heating schemes in a number of cases, for which our host country is a good example.

Spain started a cogeneration programme in 1986 aimed at an additional 700 MW electric potential to be installed in suitable industrial plants by 1992. This will result in annual energy savings of half a million TEP. Whereas direct subsidies were offered in a first phase, the programme now comprises the following activities:

- feasibility studies and co-financing of economic viability studies;

- technical aid and financial assistance for a project by third party financing and soft loans;

- information service (successful projects, most appropriate solutions to typical problems, etc.).

However, in spite of spectacular progress made up to now, a lot of work still needs to be done, not at least where the technology of cogeneration is concerned. Therefore, innovative cogeneration projects have always been eligible for financial aid in the framework of the successive energy technology programmes known as "demonstration programmes". The new THERMIE programme, proposed by the Commission and currently being discussed by the European Parliament and the Council, will continue to provide assistance for innovation in the field of rational use of energy.

Finally, the Commission invited the Council to endorse a recommendation concerning the private generation of electricity and this clearly indicates the Commission's support for the promotion of combined heat and power production. It was stated - inter alia - that:

- combined heat and power generation (CHP) and waste energy (combustion of waste and use of residual heat in industry), with their potential for oil substitution and savings of exhaustible primary energy sources, could make an important contribution to the achievement of the Community's 1995 energy policy objectives;

- the generation of electricity is an important field of application for CHP and is therefore of crucial importance to the development of this energy supply potential.

For all these reasons, the Community will continue to support the CHP technology, and the results of the next two days will certainly be a great help in this way.

OVERVIEW OF TECHNOLOGIES

Chairman : Mr. J. Sirchis

Directorate-General for Energy

Commission of the European Communities

COGENERATION TECHNOLOGIES : PRESENT AND FUTURE DEVELOPMENTS

F. ALBISU

Sener, Ingeniería y Sistemas S.A.
Bilbao-Madrid
Spain.

SUMMARY

This is an overall review of the principles on which the interest in and possibilities of cogeneration are based. Tendencies, alternatives and main comparative results are discussed after a brief introduction to efficency definitions. Apart from equipment developments and besides new cogeneration schemes on a case by case basis, two particular aproaches are mentioned: The use as heat sources for cogeneration of municipal solid waste incineration boilers or on a completely different level, of small compact, inherently safe nuclear reactors.

RESUMEN

Se exponen de forma global los principios que respaldan el interés y las posibilidades de la cogeneración. Tras una breve introducción de los tipos de rendimiento a considerar, se exponen tendencias, alternativas y principales resultados comparados. Aparte del desarrollo de equipos y de nuevos esquemas de cogeneración caso por caso, se mencionan dos sistemas concretos : El empleo como fuentes térmicas para cogeneración de calderas de incineración de residuos sólidos urbanos o, como alternativa completamente diferente, de pequeños reactores inherentemente seguros.

COGENERATION TECHNOLOGIES: PRESENT AND FUTURE DEVELOPMENTS

F. Albisu
SENER, INGENIERIA Y SISTEMAS, S.A.
Bilbao-Madrid, Spain

1. INTRODUCTION

The oil crisis of the early seventies and the subsequent increases in the price of conventional fuels prompted utilities, industry and public to look back to concepts which a cheap and apparently permanent availability of energy sources had almost made them forget: energy savings, increases in conversion efficiency, tapping of unconventional energy sources, etc. Cogeneration belongs to the same group of concepts.

Since the start of the industrial revolution (whenever it may have taken place in the different areas of the world), it was obvious that almost all industrial processes required supply of some type of energy, for tasks such as heating, drying, moving materials, etc.; throughout history, different kinds of fuel materials were assigned to meet such energy needs.

Electricity came somewhat later. Industries with high electricity needs installed power plants of their own, as an alternative to electricity purchases; these industries, in general on need at the same time of energy in the form of heat or steam, became thus self-producers of different energy forms.

Cogeneration, the simultaneous (or shared) production in a single facility of mechanical energy (usually applied to electricity production) and heat (frequently in the form of steam), was already applied in industry several decades ago, but the drastic oil price rises of the 70's and the energy crisis that followed opened the way to a keener interest in cogeneration schemes on the part of industry, government, energy agencies, etc.

A parallel effort by manufacturers to make available a full range of efficient and reliable equipment for different cogeneration situations, and legal provisions in some countries to stimulate sales of electricity by individual producers to the public networks, complete the picture in which cogeneration projects can be evaluated and later implemented.

That interest and these efforts result presently in an important increase in cogeneration-related activities, both institutional and industrial, in Spain and in the EEC countries as well as in many other parts of the world; this Seminar is a proof of it.

In the particular case of some countries under development, a frequently additional point of interest in cogeneration is the lack of a reliable nation-wide electrical grid, which forces industry to look for ways to become self-producers. This additional aspect is certainly not the case in the developed countries.

2. SOME GENERALITIES

Perhaps is not inappropriate, in the initial session of this Seminar, to introduce some of the concepts in which cogeneration is based, as an alternative to the conventional way of supplying energy to an industrial plant by power purchases from the electrical utility and fuel purchases to satisfy heat requirements.

Electricity purchased from the network comes in most cases (hydro power is the almost sole exception) from the transformation of heat; conversion efficiency at the plant outlet is some 30-35%, which drops by about five points when transmission and distribution losses are considered.

Autogeneration (or self-generation) means in-plant production of electricity (or mechanical energy) by the user; it may use different options like coal, oil, gas, hydro, diesel, etc., generating plants, up to and including wind plants, etc.

Cogeneration, on the other hand, was defined above as the simultaneous production of electricity (or mechanical energy) and heat starting from a single fuel. In most cases, a cogeneration facility produces the total plant heat requirements plus electricity which, complemented if necessary by purchases from the network, is also consumed in the plant; alternatively, electricity produced in excess of plant requirements is sold to the network. Either of these situations can be permanent or interchangeable daily, seasonally, etc. A cogeneration installation at an industrial plant will reduce transmission losses, and make heat usable that would otherwise be lost. Overall efficiency can reach 90%.

The figure shows two options for an industrial plant requiring electricity and heat in quantities E_1 and Q_1. In the first option, E_1 and the fuel to produce Q_1 units of heat are purchased.

In the second option, a cogeneration facility, supplied with purchased fuel, produces amounts E and Q of electricity and heat. Additional quantities of electricity and fuel may be purchased to satisfy the requirements E_1 and Q_1 of the industrial plant; in some situations one of those quantities, perhaps both, can be zero.

It is not simple to quantify the concept in terms of efficiency, because of the coexistence of two forms of energy, namely electricity and process heat, of so different quality.

Two efficiency figures are frequently used for a cogeneration plant. The first one is the overall efficiency already mentioned, i.e. the ratio between useful energy obtained (electricity plus heat) and energy in the fuel supplying the plant:

$$\eta = \frac{Q + E}{Q_o}$$

Here Q_o is the heat content of the fuel, and Q and E the usable heat and electrical power obtained, all three magnitudes in the same units.

The other way to quantify efficiency, perhaps more appropriate in the case of a cogeneration plant, can be called "incremental electrical efficiency". It shows the conversion efficiency (into electricity) of the heat excluding the process heat. Its expression, with the same symbols, is:

$$\eta_E = \frac{E}{Q_o - Q/\eta_s}$$

CONVENTIONAL

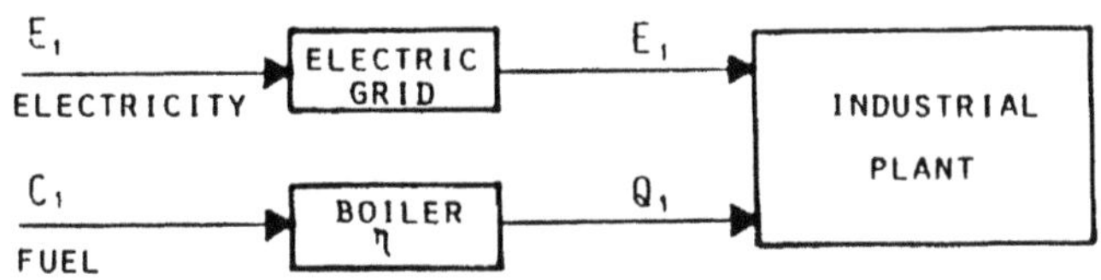

COGENERATION

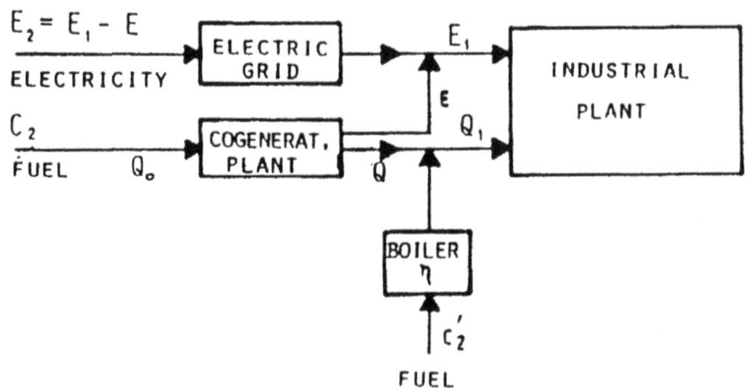

η_s is a conventional figure (which can be around 0.9) for the thermal efficiency in steam production with an ordinary boiler.

The two efficiency figures depend of course on the facility under study and, in general, on the ratio E/Q, on the process steam conditions, on the existence (or not) of post-combustion, etc. It is known (depending of the type of cogeneration) that the overall efficiency can reach up to 0.9 or more, while the incremental electrical efficiency may be between 0.5 and 0.8.

Needless to say, cases specially attractive are those where existing subproducts and wastes can be burned (whether wood chips, paper mill wastes, refinery gas, etc.), substituting totally or partially for purchased fuel.

With heat provided by the input fuel as the primary energy source for a cogeneration facility, two main families of solutions can be envisaged:

- Topping systems
- Bottoming systems

In the topping systems, heat generated by the fuel is first used to produce electricity (through a motor/generator set); afterwards, the thermal energy at lower temperature is used to produce process steam.

In the bottoming systems, the heat from burning the fuel is first used to satisfy process thermal needs; residual heat is used to produce electricity.

The topping systems are much more common since industrial processes usually require thermal energy at medium or low temperature. The bottoming processes are used only in very specific industries with processes at high temperature (for example, for heat treatments); the residual low temperature heat may be used as input source for a steam production installation, with water or organic fluids.

3. INDUSTRIAL REALIZATIONS

Very different cogeneration schemes can be contemplated after looking at the major choices at hand:

a. The primary energy source
 . Coal
 . Liquid fuels
 . Gas (natural or not)
 . Other
b. The driving engine(s)
 . Steam turbine, condensing on back-pressure, with or without extraction
 . Gas turbine
 . Reciprocating engine
 . Gas and steam turbines (combined cycle)
c. The use of mechanical energy
 . Electrical generator
 . Mechanical devices (pumps, compressors, etc.)
c. The use of thermal energy
 . Conventional boiler
 . Heat recovery boiler
 . Drying facility
 . Heating, air conditioning
 . Other

The cogeneration facility is in general better identified by the type of driving engine employed, whose selection is based on a variety of factors, including the following:

- Ratio of electricity/heat requirements
- Available fuel
- Required temperature range
- Size of facility
- New vs. old industrial plant
- Daily, weekly work schedule
- Location and, in general, environmental requirements

It is clear that cogeneration installations adopt many different solutions to satisfy electricity and heat demands of the industrial plant. The aim in each case is to achieve economy and reliability.

3.1 Cogeneration with steam turbine

This type of cogeneration system is very common in sectors such as the pulp and paper industry. Its technology has developed very consistently along the years, evolving into reliable, easy to operate, high efficiency installations.

The steam turbine, in itself a simpler equipment than the gas turbine, entails however a larger and more complex installation: boiler, pumps, water treatment, etc. Two types of steam turbines can be found, depending on the steam outlet pressure: condensing turbines and back-pressure turbines.

where gas turbines find application as prime component for cogeneration plants.

It is important to point out at the same time that gas turbines can burn not only gas, but also liquid fuels such as light oils and low sulphur fuel-oil; these uses require strict filtering systems and a higher level of equipment maintenance. Additives can help in eliminating corrosion by vanadium in the fuel.

Industrial gas turbine development has been aided by the effort put into aircraft engines. Today there are two families of gas turbines for industrial applications: the industrial type, very robust, and the aircraft-derived type, lighter and in general for lower power levels. Efficiencies (mechanical output/heat input) range between 20 and 35% with the larger values for the high power units.

The turbine drives a generator, and the hot exhaust gases (at some 500°C and with high oxygen content) make possible direct process applications (in furnaces, dryers, etc.), steam production in a heat recovery boiler (with or without post-combustion), or steam production in a conventionally fired boiler, where the gas turbine exhaust flow acts as the comburent. The post-combustion helps to tailor steam production to demand in the system with heat recovery boiler.

Both turbine families can have intermediate steam extraction, making available process steam at various conditions. Electrical production efficiencies for these turbines may vary from some 36-40% for the condensing turbines down to about one half of this value for back-pressure turbines.

In general, the system using back-pressure turbine is rather rigid in the ratio electricity/steam, making it insufficiently flexible for users with large energy variations. Variation of steam flow is easier to achieve using condensing turbine. Hence it is easier to vary broadly consumption of steam and electricity.

In many countries the availability of natural gas, which lends itself better to cogeneration (perhaps with post-combustion), makes the steam turbine less than ideal as prime element. But if natural gas is not available, or if existing oil-fired boilers are used in cogeneration, then steam turbines have their rationale. They furthermore make it realistic to consider fuels such as coal and wastes.

There is today and ample supply of coal from many regions. Although coal fired cogeneration plants require more investment and produce more pollution than other systems, oil-to-coal conversion of existing boilers offers promise for continued cogeneration. For the future (as in coal-fired utility power plants), coal-based cogeneration will rely on advanced techniques: immediately fluidized-bed combustion, and later coal gasification, the latter allowing also a shift to gas turbines.

3.2 Cogeneration with gas turbine

Natural gas has been used for a long time in many parts of the world. Europe is being covered by a gas network which will soon extend from the Urals to Lisbon, with input from the continent's own resources and from outside suppliers. Its use for power generation in central stations is certainly restricted in favour of other fuels, but its use in industry is widely promoted by governments. This is the main field

3.3 Combined cycles

In the so-called combined (gas-steam) cycle, both a gas turbine an a steam turbine drive generators producing electric power. The second turbine is driven by steam produced in the heat recovery boiler, whose primary side receives the exhaust gas from the gas turbine. For the cases where process steam is also needed, the steam turbine is of the back-pressure type; a post-combustion system may also be incorporated allowing a complete matching of production and demand for both power and process steam.

It may be added that combined cycle plants, with powers up to 100 or 200 MW per unit, are common in areas with gas supply as small central power stations, with or without process steam production.

3.4 Cogeneration with reciprocating engine

Reciprocating engines, diesel or otherwise, can also be used as main equipment for cogeneration plants. Conceptually the system would not differ very much from those based on gas turbines; there is however a substantial difference in that the thermal energy recovered from the alternating engine is at much lower temperature.

Reciprocating engines for cogeneration use various liquid fuels, and also gas. The industrial experience with these equipment has led to very robust engines, with design operating lifes of some 50,000-60,000 hours; engines derived from the automobile industry result in equipment with more favourable prices but with lower life expectancy, partly due to their higher rotation speed.

For these systems, heat recovery sources are the exhaust gases and the engine cooling system. The low temperature level of these sources restricts its application to hot, pressurized water or to low-pressure steam.

Industrial equipment falling within this concept is in general for electrical powers from some kW up to 2 or 3 MW, and with a high electricity/ heat ratio. It is very suitable for application in the tertiary sector as large hospitals, sports centres, commercial buildings, etc.

TABLE 1
COMPARATIVE PRIMARY ENERGY SAVINGS

SYSTEM		ENERGY PRODUCED kWh	COGENERATION		CONVENTIONAL		SAVINGS % OF PRIM. ENERGY
			OVERALL EFFICIENCY	PRIMARY ENERGY CONSUMPTION kWh	EFFICIENCY	PRIMARY ENERGY CONSUMPTION kWh	
Gas turbine	Electricity Heat	26.5 54.4	- -	- -	0.33 0.93	80.3 58.5	-
	TOTAL	80.9	81 %	100	-	138.8	28
Gas turbine with post-combust.	Electricity Heat	17 69.7	- -	- -	0.33 0.93	51.5 74.9	- -
	TOTAL	86.7	87 %	100	-	126.4	21
Steam turbine	Electricity Heat	15 75	- -	- -	0.33 0.93	45.5 80.6	- -
	TOTAL	90	90 %	100	-	126.1	21
Reci-proca-ting engine	Electricity Heat	36 34	- -	- -	0.33 0.93	109 36.6	- -
	TOTAL	70	70 %	100	-	145.6	31

In the lower power range packages are available incorporating engine, electric generator and heat recovery system:

3.5 Some typical energy savings

Table I shows typical results for some of the most frequent cogeneration schemes, assuming for simplicity 100 kWh of primary energy consumption in all cases. Primary energy consumptions to meet separately the power and heat requirements without cogeneration are also given. It may be seen that the system with reciprocating engine leads in energy savings, followed by that with gas turbine. The value 0.93 has been adopted as typical efficiency for the conventional transformation of the fuel energy content into steam. Note flexibility, for a gas turbine cogeneration plant, arising from use of a post-combustion system, which makes design possible in accordance with almost any heat/electricity ratio desired.

3.6 Trends in options

With circumstances so different in the existing industries and with the different alternatives offered for cogeneration, it is difficult to give standard solutions valid universally. However, some general aspects can be discussed.

First of all, it has to be stressed that only users of energy in the form of heat can opt for cogeneration. They should require large amounts of heat from hot gases or from medium to low pressure steam. The lower the process temperature the better the possibilities of cogeneration. They should furthermore have at hand a good, reliable, low-priced fuel that is not likely to become unavailable.

Each of the cogeneration solutions above has its own characteristics, which make it especially suited to some particular industry. The following general conclusions can be drawn:

- Industries burning mainly coal or wastes will most likely opt for systems based on the steam turbine. Manufacturers offer a wide power range starting at 200 kW. The steam turbine is less efficient, but investment required is lower. Furthermore the technology, operation and maintenance are simple.
- Gas if available will make it realistic to install either gas turbines or gas reciprocating engines. There is also a wide spectrum of power levels offered, from 300 kW to 200 MW for gas turbines, and from 15 kW to 2 MW for reciprocating engines (gas-fueled or otherwise).

In many countries, as in the case of Spain, regulations have been introduced that let the individual producer sell his excess electricity to the grid. He will generally find the prices offered very attractive. This situation has given a boost to the installation of cogeneration facilities.

Although most of the foregoing has been referred to cogeneration in industry, non-industrial sectors offer also a field of interest for cogeneration, with main applications for district heating and for large consumers of heat/refrigeration and of electricity: hospitals, universities, etc. District heating in particular, a field limited geographically to some areas of the world, specially the northern parts of Europe and of the American hemisphere, is looking to cogeneration as one of its alternatives.

4. SOME COMING DEVELOPMENTS

As will be seen later in this Seminar when reviewing individual projects, each plant owner will try to optimize his energy bills by selecting his best choice among the options available to him; to optimize may mean in some cases going as far as inverting the direction of the bills.

Looking at the major choices indicated at the beginning of paragraph 3 above, there is a continuous expansion of all of them: new energy sources, improved equipment, new applications for the mechanical and thermal energies produced. I am sure that much of this will be heard later today and tomorrow.

I will only mention briefly two less-than-usual energy sources and their implications.

The first one is the energy provided by burning municipal solid wastes. Certainly it is not a novelty, at least in several European countries; in Spain is a subject of growing interest within the Administration and at the large urban areas around the major cities, where the wastes represent a serious problem. Incineration seems the best choice, ahead of landfilling, composting, etc.; and here is where cogeneration comes in, since those urban areas have a great potential for consumption of steam, hot water, etc. Recent developments in MSW plants allow for a substantial increase in the power produced with a minimal increase in investment.

With MSW or with uranium, and with far less exciting energy sources in between, cogeneration will spread across the industry with benefits for everybody: the Administration, the owner and the public.

The second subject is the use of a small nuclear reactor as heat source. Aside the use of large reactors in utility power plants, small, easy to operate, inherently safe reactors of thermal powers of 100 to 500 MW offer (certainly on paper so far) excellent prospects for economical, pollution-free cogeneration; I have to say that they are not, at least today, applicable to plants with electric output of less than 50 to 100 MW. Some of these reactor designs have already left the drawing table for immediate implementation.

COGENERATION AND WOOD/BIOMASS FUELED POWER SISTEMS

ELIAS P. GYFTOPOULOS

Massachusetts Institute of Technology
Cambridge, Massachusetts
U.S.A.

SUMMARY

The purpose of this paper is to describe a number of recently installed cogeneration systems and wood/biomass fuelled power systems. Cogeneration affords one of the largest opportunities for saving fuel because many common processes have sizeable waste energies suitable for this technology. Some of the energy conversion devices, such as steam turbines and reciprocating diesel and spark-ignition engines, have been in common use for decades. Others, such as turbines with organic material as a working fluid and thermionic converters are just now being commercialized or are still undergoing testing. A survey of typical applications is presented with special references to wood/biomass fuelled power systems.

RESUMEN

El objeto de esta ponencia es la descripción de algunos sistemas de cogeneración recientemente instalados y sistemas que utilizan la madera/biomasa como combustible. La cogeneración es un sistema que ofrece una de las mayores posibilidades para el ahorro de energía dado que muchos procesos liberan energía que se puede aprovechar con esta tecnología. Algunas de las unidades de conversión de energía tales como turbinas de gas, motores alternativos diesel o motores de explosión se llevan utilizando desde hace tiempo. Otros como las turbinas que utilizan materia orgánica como combustibles o los convertidores termoiónicos se están comercializando en la actualidad o están en fase de experimentación. Se presenta un conjunto de aplicaciones típicas en este campo con especial énfasis en sistemas cuyo combustible es la madera o la biomasa.

COGENERATION AND WOOD/BIOMASS FUELED POWER SYSTEMS

ELIAS P. GYFTOPOULOS

Massachusetts Institute of Technology
Departments of Mechanical and Nuclear Engineering
Room 24-109
77 Massachusetts Avenue
Cambridge, Massachusetts 02139, U.S.A.

1. INTRODUCTION

The purpose of this paper is to describe a number of recently installed cogeneration systems and wood/biomass fueled power systems.

As it is well known, the term cogeneration refers to the concurrent generation of motive power or electricity and process heat or steam. Cogeneration saves fuel because either waste energy from a heating process is used for the generation of motive power, or waste energy from a power plant is used for heating applications. Typical fuel savings are illustrated schematically in Figures 1 and 2. For example, the top of Figure 1 shows the fuel consumption--2.25 barrels of oil (14.2 MJ)--of a high temperature heating process requiring 5.4 million British thermal units of net process heat (5.7 MJ), and the fuel consumption--1 barrel of oil (6.3 MJ)--of a power plant generating 600 kilowatt-hours of electricity. The bottom of the figure shows that the same energy services can be provided by using only 2.25 barrels of oil (14.2 MJ) to fire the high temperature process, and then capturing the waste energy from this process to supply the power plant. Thus, an energy saving of 31 percent is achieved.

Again, the top of Figure 2 shows the fuel consumption--1.75 barrels of oil (11.1 MJ)--of a low-pressure steam boiler that raises 8,500 pounds (3,860 kg) of process steam, and the fuel consumption--1 barrel of oil (6.3 MJ)--required for 600 kilowatt-hours of electricity. The bottom of the figure shows how the same energy services can be provided using only 2.25 barrels of oil (14.2 MJ). This energy is used in a boiler to raise high-pressure steam, which in turn flows into a back-pressure turbine. The turbine powers the generator, and supplies low pressure steam to the process. Here, the energy saving is 19 percent.

Cogeneration affords one of the largest opportunities for saving fuel because many common processes have sizeable waste energies suitable for this technology. It encompasses many different energy recovery and energy conversion devices. Some of the energy conversion devices, such as steam turbines and reciprocating diesel and spark-ignition engines, have been in common use for decades. Others, such as turbines with an organic material as a working fluid and thermionic converters, are just now being commercialized or are still undergoing testing. The various conversion technologies currently available and those soon to enter the marketplace provide power system designers and utility managers with an unprecedented opportunity to save not only energy but scarce capital as well.

Small-scale cogeneration facilities save capital because the equipment is built in a manufacturing plant rather than at the site of the facility, and in a much shorter time than that required for a large central electric power station. This latter feature is an invaluable tool for electric utility planners who have had to predict under conditions of great uncertainty electricity demands a decade before a new large power plant would finally come into service.

Power devices for cogeneration fall into two distinct classes: topping units and bottoming units. Topping units take advantage of the fact that many low-temperature direct-fired processes such as drying, curing, baking, space heating, and washing are thermodynamically inefficient because they consume directly the high-quality energy of high-temperature combustion products for tasks that actually require only low-quality energy. The effectiveness of fuel use in such processes can be increased substantially by first using the high-quality energy of fuel combustion in a diesel engine, gas turbine, or steam turbine to drive an electric generator, and then recovering the exhaust energy of the unit to perform heating tasks needing temperatures of only 70 to 350°C. Bottoming units are applicable to high-temperature processes such as the production of metals and ceramics in furnaces and kilns operating at 500°C and above. Waste energy from such a process is directed to a power conversion device driving an electrical generator. In a typical application, furnace exhaust gas, still containing a large quantity of high-quality energy, is directed to a boiler where steam is generated. The steam drives a turbine-generator engine and produces electricity. The combined system uses about 30 percent less energy than when the furnace heat and electricity are produced separately. Cogeneration by means of waste energy recovery with a bottoming engine is particularly attractive because it produces electricity with no incremental consumption of fuel and often can be installed in existing facilities.

Another source of cost-effective contributions to a nation's energy needs is through use of biomass either in cogeneration or power plants. Forests are one of the most valuable and renewable resources. Wood wastes generated from forest management techniques and by-products from wood processing operations can fuel electricity plants. Agricultural wastes in the form of field crop residues, tree and vineyard prunings, shells, pits, hulls, and other general processing waste are also suitable fuels for electricity generators.

2. TECHNOLOGIES

The major energy conversion technologies used in cogeneration are described briefly in what follows.

Steam Turbines. Steam turbines have been used for both cogeneration and conventional power generation throughout much of this century. In a paper mill, for example, a high-pressure topping turbine extracts part of the energy from a high-pressure steam flow. The remaining energy in the exhaust steam, at pressures of 3 to 15 atmospheres, is used to operate paper mill machinery such as digesters, blenders, and dryers. A typical electrical output would be about 50 kilowatt-hours per million kilojoules of steam energy delivered to the mill machinery.

In a district heating installation, waste energy from a power plant is fed, either in the form of low-pressure steam or hot water, to a network that supplies the heating needs of a city or a residential and commercial complex of buildings.

Low-pressure steam turbines are used as bottoming units. They recover waste energy from relatively high-temperature exhaust gases of a process by means of a waste-heat boiler, or from the spent steam of intermediate-temperature industrial processes.

Steam topping and bottoming turbines are feasible from about 2 megawatts up to several hundred megawatts with presently available hardware. Capital and installation costs for such units range from abut $1000 to $2000 per kilowatt, depending upon system size, waste energy temperature, type of fuel, and specific interface requirements and site constraints for the cogeneration system.

For district heating applications, the capital and installation costs are dictated by the type of plant under consideration and the costs of the district heating network.

Diesel Engines. Diesel engines are applicable as topping units of cogeneration systems when a high ratio of electrical output to process heat is required--up to 400 kilowatt-hours per million kilojoules of heat delivered to the process. Process steam and hot water are produced by recovery boilers coupled to the exhaust stack and to the cooling water of the engine. Systems from as little as 100 kilowatts to several thousand kilowatts can be built. However, these systems are based upon medium-speed and high-speed diesel engines, the type generally used in trucks, construction equipment, and rail locomotives. Such engines are limited to the burning of high-grade distillate petroleum, a product that is likely to be expensive and often in short supply in years to come.

A more versatile diesel engine for topping large cogeneration systems, from several thousand kilowatts up to about 30,000 kilowatts, is the large slow-speed, two-stroke diesel engine. This engine, often used for propulsion of large ships, is capable of burning very-low-grade fuels such as high-sulfur crude or heavy residual oil. Recent experiments have shown that it may even be capable of burning a powdered coal-water slurry. System costs, including heat recovery boilers, range from about $1200 to $1800 per kilowatt.

Combustion Gas Turbines. Combustion gas turbines are well suited as topping units for large-scale systems, particularly where natural gas or clean burning byproduct fuels such as refinery gas are available. Gas turbine systems offer low capital cost, about $500-$1000 per kilowatt, particularly in large systems of 10 to 150 megawatts. Also, the high exhaust gas temperature of gas turbines permits their integration with a great variety of industrial processes.

Spark-Ignition Engines. Spark-ignition engines that burn natural gas can also be used as topping units. A relatively new concept for achieving very low capital cost is based upon derated automobile engines converted for use in prepackaged cogeneration modules. One module generates about 30 kilowatts of electricity and about 230,000 kilojoules per hour of hot water at 110°C. Another module generates about 60 kilowatts of electricity and about 460,000 kilojoules per hour of process heat in the form of low pressure steam and hot water. Combinations of several modules can be used in applications such as shopping centers, hospitals, apartment buildings, and light industrial sites, to supply all on-site electrical and process heat needs.

Other modules are rated at 200 kilowatts, and 600 kilowatts of electricity, and proportionately higher thermal outputs, including relatively high pressure steam. For example, a natural-gas, turbocharged internal combustion engine, coupled with an electric generator and a twin-helical screw steam compressor can generate between 480 and 650 kilowatts of electricity, and between 1400 and 1700 kilograms per hour of high pressure process steam at about 10 atmospheres. Prior to the introduction of the screw compressor, cogenerators requiring high-pressure process steam were forced to use combustion turbines rather than reciprocating engines which yield much higher electrical output efficiency.

Organic Rankine Turbines. An organic Rankine turbine is an advanced type of bottoming unit. It uses an organic material as a working fluid and is capable of recovering efficiently the energy from low-temperature (150 to 600°C) waste streams. It can be built in a wide range of sizes, from as small as 50 kilowatts to 30,000 kilowatts or more. Output per unit of waste energy input will generally be 20 to 30 percent greater than that obtainable with steam-turbine bottoming units. Commercialization of organic Rankine turbines is just beginning.

The various technologies described above provide the basis for virtually all cogeneration systems. Other technologies now in the research and development stage will also play a role in future cogeneration systems.

3. TYPICAL APPLICATIONS

Cogeneration

Cogeneration has been practiced for many decades. The advent of the energy crisis in the 1970's rekindled the interest in cost-effective, energy-saving technologies, in general, and cogeneration in particular. A few examples of recent additions to the U.S. cogeneration capacity are as follows.

A number of units have been developed, and are owned and operated by Applied Energy Services. One of these is a $280 million petroleum coke-fired facility in Houston, Texas, designed and constructed by Bechtel Power Corporation. Its electrical rating is 140 megawatts, and its thermal output is 15 short tons of process steam per hour. The electricity is sold to the Houston Light and Power Company, and the steam to the local ARCO refinery which also supplies the petroleum coke. The plant began commercial operations in July 1986.

Another unit is a $116 million coal-fired plant purchased from ARCO, and refurbished by Bechtel. It is located in Monaca, Pennsylvania. It generates 121 megawatts of electricity, and 43 short tons of process steam per hour. The electricity is sold to West Penn Power, and the steam to ARCO Chemical. The plant became operational in July 1987.

A third plant is a $120 million gas-turbine project in Newhall, California, designed and constructed by Brown Boveri Corporation. It generates about 100 megawatts of electricity sold to Southern California Edison, and 125 short tons of process steam per hour supplied to local oil leases and other steam users. It began operations in 1988.

Many smaller cogeneration plants have been designed and built by Thermo Electron Corporation.

One is a diesel cogeneration system at the Hoffmann-La Roche chemical plant in Belvidere, New Jersey. It generates 23 megawatts of electricity, and can also produce 72.6 tonnes of process steam per hour, and 119 tonnes of 76.6°C water per hour. It supplies all the electrical and thermal needs of the chemical plant, and excess electricity is sold to the local utility. The plant began commercial operation in December 1982, and achieves the overall energy use of 87 percent. Without cogeneration, the energy consumption would have been larger by the equivalent of 200,000 barrels of oil per year.

A simple schematic of the Hoffmann-La Roche plant is shown in Figure 3. The engine is a 10-cylinder Sulzer 10 RNF 90 M, two-stroke diesel which delivers 23.3 MW at 120 r/min. It has a 900 mm bore and 1,550 mm stroke. Overall height is 11.6 m with a baseplate of 4 m and a length of 21.51 m. Net weight is 980 tonnes. It operates on residual fuel.

The generator is manufactured by Siemens, and is a 60-pole, three phase, 13,800 volts, 60 Hertz synchronous unit.

Waste heat from the diesel engine is recovered from the exhaust gases, air cooler, and engine water cooling circuits. In order to maximize the overall thermal efficiency, the temperature levels of the waste heat are matched to the plant thermal requirements.

The boiler is supplementary fired because the chemical plant has a requirement of up to 72.6 tonnes of 15 bar steam, much greater than the amount that can be obtained without the supplementary firing. Additional oxygen beyond that already contained in the gases is not necessary because of the large amount of excess air used in the two-stroke diesel engine.

An energy balance of the plant is shown in Figure 4.

A second example is an installation at the downtown Government Center in Dade County, Florida. This cogeneration unit began operation in December 1986. All of the electrical power, air-conditioning, and hot water needs of the Center are met by a $30 million combined-cycle cogeneration system supplied on a turnkey basis by Thermo Electron.

The Dade County Downtown Government Center is a complex of seven buildings, including a 30 storey office block, county courthouse, public library, museum, and a center for the fine arts. The cogeneration system installed meets the electricity, air conditioning, and hot water needs of the complex with an energy efficiency in excess of 76 percent when the air conditioning load is highest.

At the heart of the system are two turbine generator sets (Figure 5). The main electricity generation is provided by a Rolls-Royce SK30 industrial Olympus gas turbine with a maximum continuous site rating of 22 MWe. Normally, the turbine operates on natural gas but it is capable of burning fuel oil in emergency or abnormal conditions. The turbine exhaust is ducted to an unfired, dual pressure, natural circulation waste heat recovery boiler providing steam for a 10 MWe Peter Brotherhood dual pressure condensing turbine. High pressure steam (42 bar) is taken from one section consisting of a superheater, steam generator, and economizer. The exhaust gases then pass through a second section consisting of another steam generator and economizer producing steam at 1.4 bar. Exhaust gas leaving the boiler is ducted to a dual-wall steel exhaust stack.

The high pressure steam is fed to the steam turbine. When power demand is high, the low pressure steam is also routed to the turbine. At times of high air conditioning demand, all low pressure steam and additional low pressure steam taken from between the low pressure and high pressure sections of the steam turbine is routed to the absorption chillers, which have a combined maximum output of 18.3 MW of refrigeration.

Condensate from the chillers is pumped through a heat exchanger before being returned to the deaerator for the production of up to 1,200 litres/min of domestic hot water.

Cogeneration modules of 30 to 600 kilowatts are manufactured by Tecogen, a majority-owned subsidiary of Thermo Electron Corporation. Modules have been installed and are being operated for a great variety of uses. A sixty kilowatt unit has been installed in each of the following sites: an athletic club in Escondido, an athletic club in San Juan Creek, the Capistrano by the Sea Hospital and Clinic, and a Ramada Inn, all in Southern California. The annual savings in each of these installations are between $20,000 and $30,000, and the payback period is between two and three years. Six Tecogen modules, 60 kilowatts each, are operating on the campus of Albion college in Michigan since December 1984. They provide electricity, hot water for showers, space heating, and swimming pool heating. Also, a four Tecogen system, rated at 240 kilowatts, is installed at a 21,200 square meter building complex in North Haven, Connecticut. The system satisfies the electricity, hot water, and heating and cooling requirements of the buildings.

A 200-kilowatt gas-fueled Tecogen module is providing electricity, space heating, and hot water to a Sheraton Hotel in Danvers, Massachusetts. A duplicate unit is operating at OK Towel and Uniform Supply, a commercial laundry in Elizabeth, New Jersey. Two 500 kilowatt units have been installed by New England Electric System at a paper mill and a tool manufacturing plant, both in Massachusetts.

Configuration schematics for the 30, 60, 200, and 600 KW modules are shown in Figures 6 to 9, and technical specifications are listed in Table 1.

Unit	CM30	CM60	CM200	CM600
Outputs:				
Electrical	30 kW	60 kW	200 kW	480 to 650 kW
Hot water	230 MJ/hr at 110°C	460 MJ/hr at 110°C	1,250 MJ/hr at 93°C	Amount depends on steam rate
Steam	---	---	---	1,700 kg/hr at 1 bar to 1,450 kg/hr at 8.5 bar
Input	Natural gas or propane			
Overall energy efficiency (%)	89.5	93.7	86.6	84
Controls	Completely automated. Unattended operation.			

Table 1: Specifications of Tecogen Modules

Wood/Biomass Fueled Systems

A number of biomass fueled electric power systems have been built by Thermo Electron. The Hemphill Power and Light project (Figure 10), in Springfield, New Hampshire, the Whitefield Power and Light project (Figure 11), in Whitefield, New Hampshire, and the Gorbell project (Figure 12), in Athens, Maine, are three wood-fueled electric power plants. Each generates 16 MW of electricity, is fueled by sawmill residue and whole tree chips, and has a cost of $\$31 \times 10^6$. The first two went into commercial operation in 1987, and the third in the summer of 1988.

Biomass fuel delivered to the plant, first passes over a weighing station and then is dumped onto the processing line. Conveyors transport the fuel to a processing facility for size separation. Fuel that is two inches or under, in all dimensions, passes through a rotating disc screen. Fuel over two inches passes to a swing hammermill for size reduction down to two inches. Fuel can then be conveyed to the boiler feed bin or moved to storage, which can be open pile, covered pile or silo.

The boiler fuel feed system incorporates live bottom surge hoppers to maintain the fuel inventory needed for operational flexibility. The steam generator is a bottom-supported, field erected, water cooled vibrating grate and balanced draft boiler. Hot gases generated in the furnace pass through the superheater, boiler bank, economizer and air heater sections before entering the flue gas cleaning system which typically consists of cyclone collectors followed by an electrostatic precipitator.

Power is generated by a single inlet, extraction/condensing steam turbine connected to a generator. The operation of the fuel processing, steam generating unit, air equipment, plus the cooling tower and electrical transmission, is controlled and monitored from a central control room.

Three agricultural waste power plants are being built in California. The Mendota Biomass Power, Ltd., in Mendota (Figure 13) is a 28 MW electric power plant using a circulating fluidized bed boiler, and fueled by woodwaste and prunings from orchards and vineyards. Its cost is $\$70 \times 10^6$. It went into commercial operation in the summer of 1989. It sells its electricity to Pacific Gas and Electric.

The Woodland Biomass Power, Ltd., in Woodland is a 28 MW electric power plant using a circulating fluidized bed boiler, and fueled by rice hulls, rice straw, orchard prunings, and woodwaste. Its cost is $\$80 \times 10^6$, and it is scheduled for commercial operation in late 1989. It will sell its electricity to Pacific Gas and Electric.

The Delano Energy Company, Inc., in Kern County is a 30 MW electric power plant using also a circulatory fluidized bed boiler, and fueled by wood and agricultural wastes. Its cost is $\$85 \times 10^6$, and it is scheduled for operation in mid-1990. It will sell its electricity to Southern California Edison Company.

The fluidized bed boilers in the three California plants are used with special flue gas treatment such as thermal de-NOx and/or baghouse to comply with the very strict environmental regulations of the State of California.

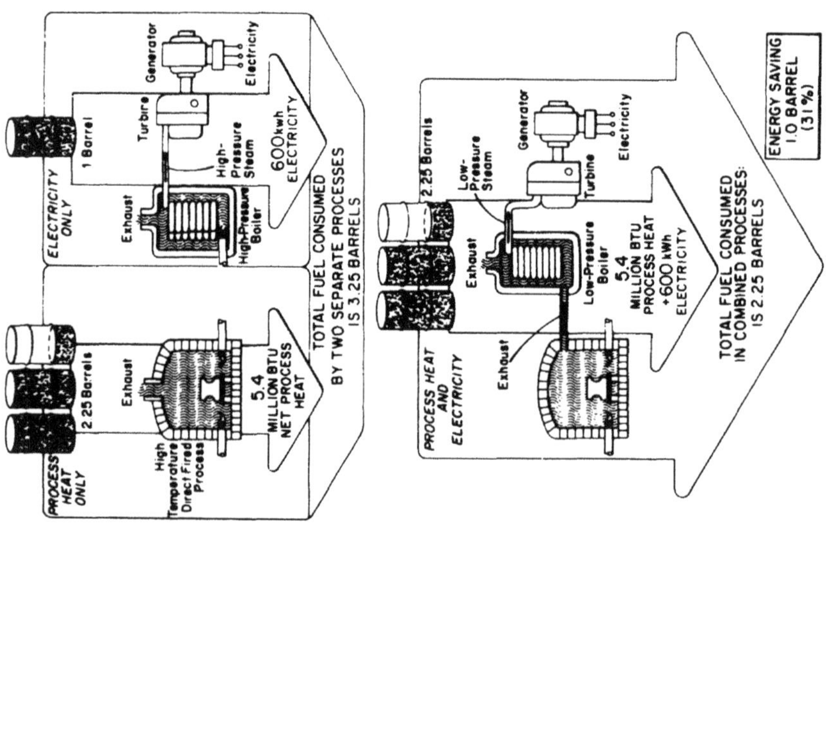
PROCESS HEAT ONLY
2.25 Barrels
Exhaust
High Temperature Direct Fired Process
5.4 MILLION BTU NET PROCESS HEAT
ELECTRICITY ONLY
1 Barrel
Turbine
Generator
Exhaust
High-Pressure Steam
High-Pressure Boiler
Electricity
600 kwh ELECTRICITY
TOTAL FUEL CONSUMED BY TWO SEPARATE PROCESSES IS 3.25 BARRELS
PROCESS HEAT AND ELECTRICITY
2.25 Barrels
Exhaust
Low-Pressure Steam
Low-Pressure Boiler
Turbine
Generator
Electricity
5.4 MILLION BTU PROCESS HEAT +600 kWh ELECTRICITY
TOTAL FUEL CONSUMED IN COMBINED PROCESSES: IS 2.25 BARRELS
ENERGY SAVING 1.0 BARREL (31%)

Fig. 1

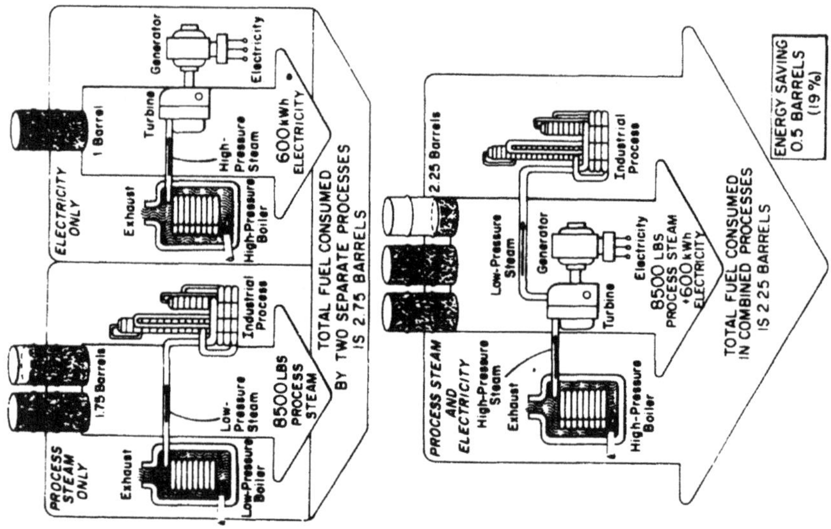
PROCESS STEAM ONLY
1.75 Barrels
Exhaust
Low-Pressure Boiler
Low-Pressure Steam
Industrial Process
8500 LBS PROCESS STEAM
ELECTRICITY ONLY
1 Barrel
Exhaust
High-Pressure Boiler
High-Pressure Steam
Turbine
Generator
Electricity
600 kWh ELECTRICITY
TOTAL FUEL CONSUMED BY TWO SEPARATE PROCESSES IS 2.75 BARRELS
PROCESS STEAM AND ELECTRICITY
2.25 Barrels
High-Pressure Steam
Exhaust
High-Pressure Boiler
Turbine
Low-Pressure Steam
Generator
Electricity
Industrial Process
8500 LBS PROCESS STEAM +600 kWh ELECTRICITY
TOTAL FUEL CONSUMED IN COMBINED PROCESSES IS 2.25 BARRELS
ENERGY SAVING 0.5 BARRELS (19%)

Fig. 2

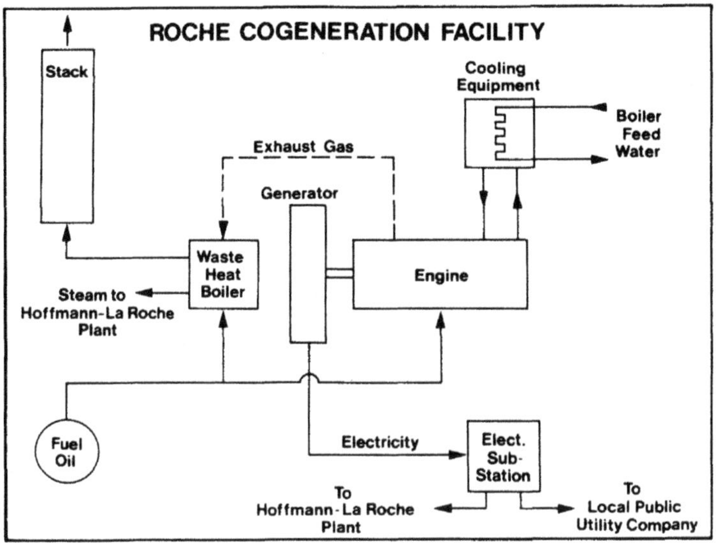

Fig. 3

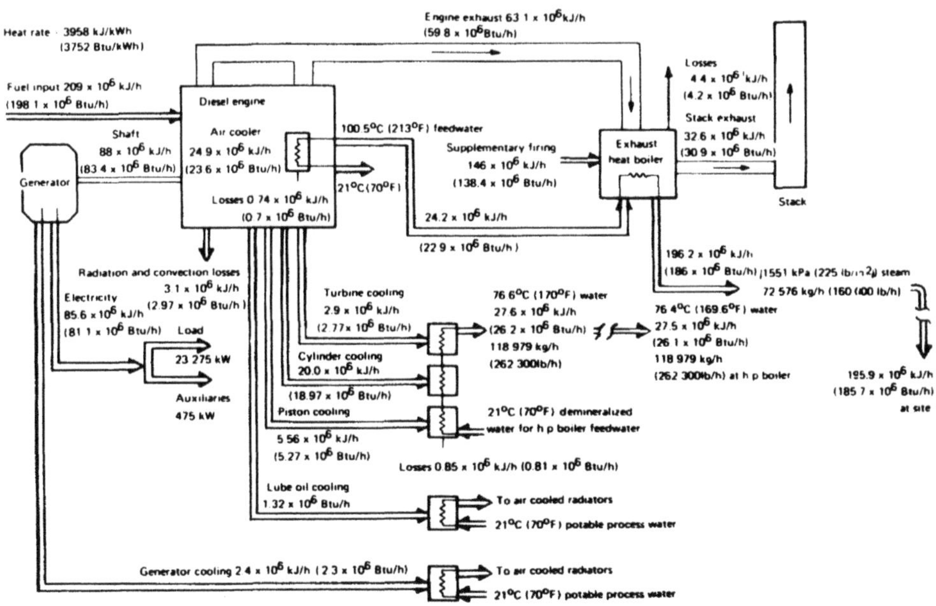

Cogeneration system energy balance (23.3 MW, ambient air at 20°C, 68°F)

Fig. 4

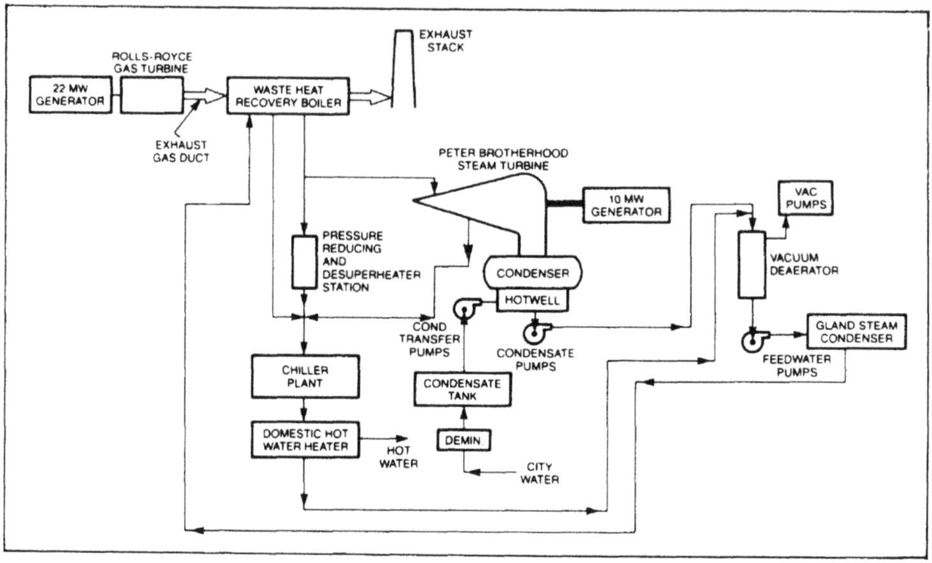
EXHAUST STACK
ROLLS-ROYCE GAS TURBINE
22 MW GENERATOR
WASTE HEAT RECOVERY BOILER
EXHAUST GAS DUCT
PETER BROTHERHOOD STEAM TURBINE
10 MW GENERATOR
VAC PUMPS
PRESSURE REDUCING AND DESUPERHEATER STATION
CONDENSER
HOTWELL
VACUUM DEAERATOR
COND TRANSFER PUMPS
CONDENSATE PUMPS
GLAND STEAM CONDENSER
FEEDWATER PUMPS
CHILLER PLANT
CONDENSATE TANK
DOMESTIC HOT WATER HEATER
HOT WATER
DEMIN.
CITY WATER

Fig. 5

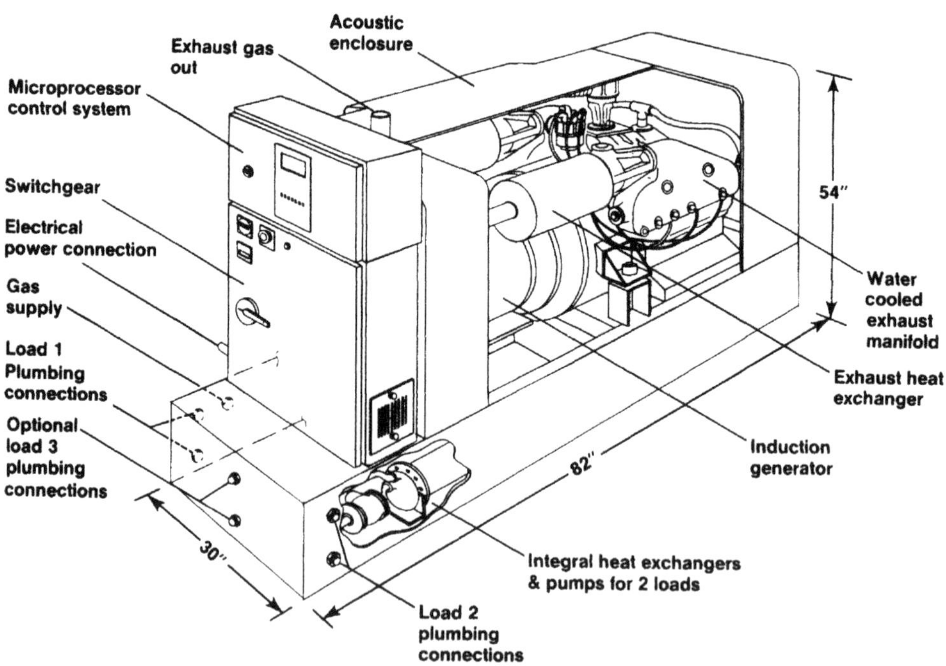
Acoustic enclosure
Exhaust gas out
Microprocessor control system
Switchgear
Electrical power connection
Gas supply
Load 1 Plumbing connections
Optional load 3 plumbing connections
54"
Water cooled exhaust manifold
Exhaust heat exchanger
Induction generator
82"
30"
Integral heat exchangers & pumps for 2 loads
Load 2 plumbing connections

Fig. 6

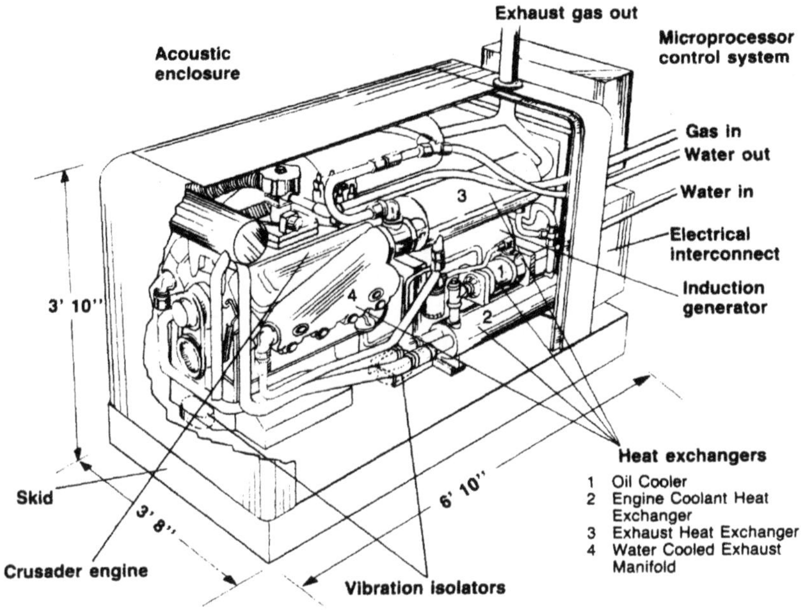
Exhaust gas out
Acoustic enclosure
Microprocessor control system
Gas in
Water out
Water in
Electrical interconnect
Induction generator
3' 10''
Skid
3' 8''
6' 10''
Crusader engine
Vibration isolators
Heat exchangers
1 Oil Cooler
2 Engine Coolant Heat Exchanger
3 Exhaust Heat Exchanger
4 Water Cooled Exhaust Manifold

Fig. 7

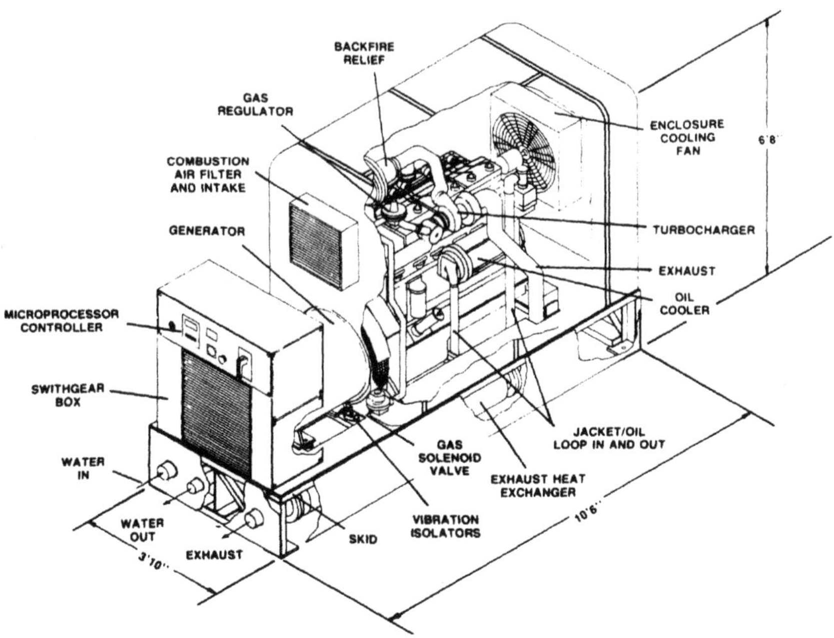
BACKFIRE RELIEF
GAS REGULATOR
COMBUSTION AIR FILTER AND INTAKE
GENERATOR
ENCLOSURE COOLING FAN
6'8''
TURBOCHARGER
EXHAUST
OIL COOLER
MICROPROCESSOR CONTROLLER
SWITHGEAR BOX
WATER IN
WATER OUT
EXHAUST
3'10''
SKID
GAS SOLENOID VALVE
VIBRATION ISOLATORS
EXHAUST HEAT EXCHANGER
JACKET/OIL LOOP IN AND OUT
10'6''

Fig. 8

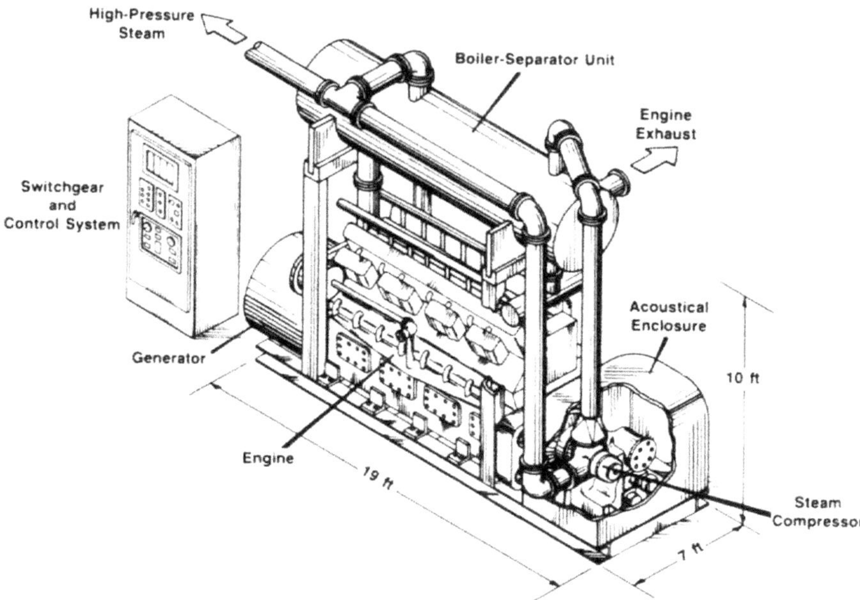
High-Pressure
Steam
Boiler-Separator Unit
Engine
Exhaust
Switchgear
and
Control System
Acoustical
Enclosure
10 ft
Generator
Engine
19 ft
Steam
Compressor
7 ft

Fig. 9

Fig. 10

Fig. 11

Fig. 12

MENDOTA BIOMASS POWER, LTD.

Fig. 13

STATE OF CO-GENERATION IN SPAIN

D. CONTRERAS, A. GOMEZ-ANGULO
Dept. for Cogeneration and Substitution
IDAE, Spain

SUMMARY

In the course of this report, a detailed analysis will be made of the present situation and recent developments in co-generation in Spanish industry.

Thus, taking as our point of departure information pertaining to 1987, the latest year for which statistics are available, an outline will firstly be given of those systems set up since then as well as of projects now in an advanced stage of construction: special features which characterize these new facilities will also be described.

Secondly, after taking the above results into account, the present state of co-generation in Spain today will be fully set out.

Lastly, an analysis will be made concerning the degree to which the potential for this technology has been tapped since being revealed through IDAE's 1987 market research into co-generation; and this in turn will enable foreseeable future development for this alternative source of energy supply to be determined.

RESUMEN

A lo largo de este artículo se efctúa un pormenorizado análisis de la situación actual y reciente evolución de la cogeneración en la industria española.

Para ello, tomando como punto de partida la información relativa a 1987, año de la última estadística disponible, en primer lugar se exponen las realizaciones de estos sistemas posteriores a esa fecha junto con los proyectos que están en fase avanzada de construcción, describiendo los aspectos especiales que caracterizan a las nuevas instalaciones.

En segundo lugar, y tras integrar los resultados anteriores, se establece lo que constituye la situación de la cogeneración en España hoy.

Por último, se analiza el grado de cumplimiento del potencial detectado de esta tecnología en el Estudio del Mercado de la cogeneración realizado por IDAE en 1987, lo que en definitiva permite definir el previsible desarrollo futuro de esta alternativa de abastecimiento energético.

STATE OF CO-GENERATION IN SPAIN

D. CONTRERAS and A. GOMEZ-ANGULO
Department for Cogeneration and Substitution
IDAE, Spain

1. ELECTRIC POWER PRODUCTION IN SPAIN IN 1987

Electric power production in Spain in 1987 was in the order of 133 390 GWh compared with 129 150 GWh for 1986, thus representing an increase of 3.3%.

Table 1 gives a breakdown of total power production by reference to source. An analysis of same shows that hydroelectric power has risen by 2.7%; likewise, thermoelectric production has gone down by 0.5% whereas nuclear-generated power has gone up by 10.2%.

2. AUTO-GENERATED ELECTRIC POWER PRODUCTION IN SPAIN IN 1987

2.1 Degree of Auto-generation and Co-generation

The industries involved in auto-generated electric power produced 4 191 GWh in 1987, which represents a 12.9% rise over the previous year (3 712 GWh).

A rise in production of this nature can be traced to an increase in thermoelectric generated power, as shown in Table 2; indeed, compared to a rise of 7.2% in auto-generated hydroelectric power, thermoelectric auto-production or co-generation, as we shall proceed to call it, went from 2 291 GWh in 1986 to 2 668 GWh for 1987, representing an increase of 16.5%.

In view of the above-mentioned figures, and as will become clear from Table 3, one can deduce that the level of co-generation in Spanish industry, as defined by the quotient between co-generated electricity and total electricity production, has experienced a 13% growth over 1987, going from 1.77% in 1986 to 2.00% by the end of 1987.

2.2 Analysis of Co-generated Power Production in the Autonomous Communities

Table 4 lists the figures for co-generated power production in the different Autonomous Communities and their respective proportional participation in the total.

Four Communities, namely Andalucía, the Canary Islands, Cantabria and Castilla-La Mancha, generated more than 50% of the total of co-generated power.

Table 5, which indicates the electrical energy co-generated on a regional scale alongside the net consumption pertaining thereto, shows on examination that, with the exception of the Canary Islands and Cantabria, the contribution of co-generation systems to total electricity consumption is very low (2.38% on average). However, as can be seen from Table 6, this parameter has risen by 12.3% over the 1986 level, owing to the fact that growth in co-generated production has been greater than the rise in consumption of electricity.

Nevertheless, in order to work with a uniform set of figures, given that net consumption excludes own consumption, consumption employed in pumping or lost in transmission and distribution, figures for co-generated production should be correspondingly reduced. Accordingly, if it is net co-generated production (i.e. gross production less own consumption) that is to be considered, the 1987 supply would then be in the order of 2.07%.

2.3 Analysis on an Industrial Sector-by-Sector Basis of Co-generated Power Production

Spanish industries engaged in co-generation can be broken down into nine sectors and a general view of their position on balance as regards electricity is given in Table 7.

The key used throughout this report is set out in Table 0.

Over 75% of total energy produced is concentrated in four industrial activities (paper, refining, steel and chemicals). The reason for this technology's high rate of acceptance among the paper sector can be put down to the very nature and needs of the paper manufacturing process. In the case of the remaining sectors, the availability of residual fuel or heat capable of use in co-generation systems, justifies the introduction of same.

If the total power needs of the co-generating sectors are taken into account, the degree of self-supply amounts to 7.3%. Since this low level refers to the global consumption of the sectors involved, it does not reflect one important feature, namely, that in industries possessing co-generation facilities, the demand for external power is in the order of 19% of total net consumption; and this is without said figures taking into account the amount of electricity channelled into the grid.

2.4 Distribution by Size of Co-generated Power Production

Table 8 provides a breakdown of total power obtained by means of systems of co-generation according to the level of production of each plant, indicating moreover the frequency of each level.

2.5 Use of Fuels in Co-generated Power Production

As will be clear from Table 9, which breaks down electricity production according to the fuel used in co-generation facilities, fuel-oil is used to produce 46% of the total.

The importance of residual fuel, with which approximately 13% of production is generated, must also be stressed.

The proportion of natural gas used in co-generating plants, though still low, has experienced a rise of 46% over 1986.

2.6 Distribution of Co-generated Power Production by Reference to Technology

Of the total 65 co-generation plants active in 1987, 54 use the back-pressure steam turbine as their generating unit and produce 85% of the total of co-generated power.

The technological breakdown for the remaining 11 is as follows: five run condensed steam turbines, three run diesel units and three run gas engines or turbines.

These data have to be understood within the context of the age of the installations in operation, with only two plants being post-1980.

The technology available today, plus the ever growing penetration of natural gas, are factors which will reverse the current ranking of co-generating systems now in use and undoubtedly give rise to a marked trend towards gas engines and turbines.

3. POWER CAPACITY OF CO-GENERATING PLANTS ACTIVE IN 1987

The power capacity possessed by co-generating plants active in 1987 is supplied in Table 10. Also appearing alongside the information concerning the previous year are the average hours of utilisation, where it can be observed that, in contrast with an increase of 13% in active power, there has been a lower rate of rise of 3% in the average utilisation to which the facilities have been put.

It is important to note that this increase in plant power capacity is not due to a greater number of plants but instead to modifications made to the operating conditions of those already in existence; as a matter of fact, in 1987 only one new plant came on line while two ceased to operate.

3.1 Analysis of Co-generated Power Capacity vis-a-vis the Autonomous Communities

The regions of Andalucía, Cantabria and Castilla-León account for almost 50% of the total active co-generated power; bearing in mind section 3.2 below it can be seen that no close relationship exists between those Communities having the greatest electricity output and those with most plant power capacity. Data for the different Communities are shown in Table 11.

3.2 Distribution on an Industrial Sector-by-Sector Basis of Co-generated Power Capacity

Table 12 gives a breakdown of active power capacity of co-generating installations according to category of industry: two sectors traditionally using this technology, namely paper and food, concentrate 53% of the total co-generated power in 43 plants.

3.3 Size of Plants

Although the average power capacity of Spanish co-generating installations is 12 MW, the majority (46 out of a total of 65) possess less.

Table 13 shows the distribution of active co-generated power capacity according to plant size.

3.4 Analysis of Co-generated Power Capacity by Reference to Technology

As already indicated when analysing electricity production according to the different systems of co-generation, the most widespread technology found in Spain in 1987 was the back-pressure steam turbine. The power capacity of these systems is 655 MW, which represents 87% of the total.

Table 14 indicates both the plant power capacity corresponding to and the number installed of the different types of generating devices.

4. NEW CO-GENERATING PLANTS

The period analysed above corresponds to the latest year for which Electric Energy Statistics are available and it is clear that, until 1987, the installation of co-generation was little in evidence.

From this date onwards, however, this technology has seen considerable development, development fundamentally due to the gap between the cost of electricity and that of fuel which allows for speedy amortisation of the investment in question; and secondly, the growing penetration of natural gas has made it possible to use engine units, such as gas turbines and engines, which possess a high degree of electrical efficiency.

This is borne out not only by the installations that have come into operation since 1987, but also by those still under construction. Before going on to outline the chief features of these plants, it should be stressed that the list given here may not be complete, in view of the fact that it has been compiled by IDAE on the basis of information received from the different agents involved in this technology.

4.1 Operational Plants

From 1987 until now, 24 co-generating projects have come into operation, having a total plant power capacity of 83 MW and an annual electric energy production of 562 646 MWh.

Table 15 summarises the fundamental aspects of these new installations, the coming into operation of which has meant an 11% rise in co-generated plant power capacity and a 21% rise in electricity generated by this technology compared to 1987.

The reason for the sharper increase in electric energy production lies in the greater number of hours during which the new plants are put to use; viz, an average of 6 773 h.p.a., compared to 3 542 h.p.a., for operational systems in 1987.

4.1.1 Analysis vis-à-vis Autonomous Communities

As can be seen from Table 15, 42% of the new installations possessing nearly 60% of the total plant power capacity and electricity production are concentrated in the Autonomous Community of Catalonia. Attention should also be drawn to the first example of this technology in Madrid and to the fact that the Valencian region has undergone a notable upswing, with five new projects increasing the present level of co-generated plant power capacity twelve-fold.

4.1.2 Industrial sector-by-sector analysis

Of 24 new plants, 14, accounting for almost 80% of total plant power capacity and electricity production, are to be found in two sectors traditionally using this technology, namely chemicals and paper.

A sector-by-sector breakdown of new co-generating projects likewise brings out the fact that this technology has been introduced as a system for supplying energy to sectors which until now had made no use of it here in Spain, sectors such as the brick and ceramic, tile, glass, timber and graphic arts industries.

4.1.3 Size of plants

The average size of plants coming on stream since 1987 has been in the order of 3 461 KW, indicating that these new schemes are on a much smaller scale than those which were in operation until said date and which, it must be recalled, had an average size of 11 589 KW, this being more than triple the average figure found at the present time.

The breakdown, in ascending order of electric power capacity, is as follows:

Number of plants	Power (MW)
8	< 1
5	1 - 3
6	3 - 4
3	4 - 6
1	8 - 10
1	> 10

4.1.4 Fuels used

As has already been pointed out when discussing fuel use in co-generated electricity production for 1987, a notable growth has been experienced in natural gas.

Proof of this resides in the fact that this fuel is employed in 17 new plants in order to generate 519 340 MWh annually, that is to say, 92% of all electricity produced by these systems post-1987; which amounts to nigh on doubling the share of natural gas as regards electric energy produced by means of this technology compared to levels recorded for 1987.

4.1.5 Technologies

A logical consequence ensuing from the above mentioned penetration of natural gas is that, in the main, the technology used to equip new facilities is that which makes use of gas turbines or engines for the purposes of generating electricity.

The number of schemes fitted out with this equipment comes to 15 (ten with single cycle gas turbines, three combined cycle and two with reciprocating engines) and together they produce 87% of all energy co-generated by the new systems. The power capacity of these schemes is 72 MW, which represents 86% of the total.

The technology employed in the nine remaining plants is based on the use of back-pressure steam turbines as the main engine.

4.2 Plants Under Construction

The 18 installations now underway and projected to come on line within the next six months will mean a rise in power capacity of 126 MW and the generation of 927 568 MWh of electric energy annually.

These figures can be seen in Table 16 with a breakdown split up according to the different criteria used for classification herein (technology, Autonomous Community, sector and fuel).

From the point of view of the siting of these new systems, mention must be made of the setting up in Andalucía of a plant having a total power capacity of 51 MW. As to sector-by-sector distribution, this technology's penetration of the chemicals and paper sectors remains steady, with a 42% share of the total projected electric capacity and production now underway, while the automobile, rubber and plastics industries appear as new users of co-generating systems.

The features of the schemes under construction and their resemblance of plants already in operation show that the development of co-generation in Spanish industry is undergoing a marked change as regards the basic parameters of such installations.

Thus, an average profile could be drawn up for these new schemes on the basis of the main units being a gas turbine with a power capacity in the order of 5 MW and intensive use being made of such facilities approaching 7 000 h.p.a.; in contrast, as has been observed above, co-generating systems set up prior to 1987 were, by and large, based on the use of back-pressure steam turbines, while average power capacity and hours of use were 12 MW and 3 542 h.p.a. respectively.

4.3 Co-generation in Spanish Industry for the Period 1988-1989

Once the schemes now under construction come on stream, the number of co-generating plants developed over the two year period 1988-89 will amount to 42, their total power capacity will be 209 MW and electricity production will reach 1 490 214 MWh annually, representing increases of 65%, 28% and 56% respectively over the 1987 figures.

To obtain an overall picture of the state of co-generation in Spain today, the figures outlined thus far must be integrated in such a way that the position existing in 1987 is seen together with the schemes which have come into operation since then and those which are still in an advanced stage of construction.

The end result of the above exercise is to be seen in the situation reflected in Table 17, which summarises the fundamental aspects characterising both the present state of affairs as well as the recent developments in co-generation in Spain, all of which has been described in the course of this report.

The chief conclusion that can be drawn is that the 1989 level of co-generation will reach 3.12%, representing a 56% growth over 1987.

It must be said that for the purposes of calculating the above figure, the 1987 figures have been kept constant with regard to total production and co-generated electric energy; we have thought it best to make no hypothesis as to trends, owing to the different rate of growth over the past few years.

The above results are ample witness to the boom that this technology has been experiencing during the past three or four years in Spain.

The explanation for this fact is closely linked with the gap between the price of electricity and that of fuels that has existed throughout this period.

Also to be kept in mind is the factor that the time required for these plants to come on stream is in the order of two years, if one includes, along with the setting-up period itself, a moderate interval for the final investment decision. Indeed, systems which have become operational in 1988-89 can trace their first beginnings back to 1986-87; and during the latter mentioned period the margin in energy prices - which in effect is what makes the amortisation of these plants possible - experienced a clearly rising trend, a trend which continued upwards until June of this year when the price of fuel was increased.

Despite the fact that the present margin might still be attractive, future movements in fuel oil prices will have to be monitored in order to make a more accurate evaluation as to whether the recent increase could lead to a slowdown in the development of co-generation in Spain.

5. ENVISAGED DEVELOPMENT OF CO-GENERATION IN SPAIN

In 1987 the Institute for Diversification and Saving of Energy (Instituto para la Diversificación y Ahorro de la Energía) (IDAE), carried out an analysis of the potential market for co-generation with the basic aim of evaluating the future of this technology in Spanish industry.

As a result of the work undertaken, the Effective Co-generation Potential was arrived at: this, depending on certain criteria of penetration being met, represented the practical application of the technologically viable potential. The most representative figures of said potential were:

No. of Installations	102
Plant Power Capacity	585 MW
Co-generated Electricity	4 321 GWh.p.a.

The timespan foreseen for converting said potential into reality was five years (1988-92). Two years have elapsed since then and in view of the activity embarked upon in this interval it would seem an opportune moment to analyse the fulfilment of said goals.

Table 18 shows the geographical and sector-by-sector distribution of the Effective Potential together with performance figures for installations post-1987.

On having completed 41% of the plants, the increase in power capacity and electric energy production amount to 36% and 34% respectively of the quantities envisaged. This means, firstly, that the average size of the schemes is slightly smaller than that defined by the market analysis; and secondly, that the co-generating systems set up are, on average, kept running fewer hours than was expected.

Analysing performance on an Autonomous Community basis, one group stands out: composed of Cantabria, Castilla-La Mancha, Navarre and Rioja, which together comprise 16% of the Effective Potential insofar as power capacity and electricity production are concerned, it has seen the installation of no new plants whatsoever. Other areas with below average achievements are Asturias, Castilla-León, Galicia, Madrid and the Basque country, the figures for this last mentioned region being of greater interest owing to the fact that it has a higher estimated potential.

Lastly, even though they together embrace 45% of Effective Potential, the development of this technology in the regions of Andalucía, Aragón, Catalonia and Valencia exceeds the overall average figure.

Taking a sector-by-sector angle, market research revealed latent potential in the steel, cement and petrochemical industries, a potential which to date has still not been realised.

Likewise, the glass, rubber, food and textile sectors have experienced a below average rate of growth. This is especially significant in the case of the food industry where, with an estimated potential of 53 MW in 14 plants, only two schemes totalling 4 MW have been completed.

The degree to which such plants have been installed in the brick and ceramic, automobile, timber, refining, chemicals and paper sectors, together representing nearly 64% of the total Potential, has been higher than expected in the time elapsed: owing to their allocated proportion of Effective Potential, the three last mentioned sectors exert a notable influence on the final figures.

The original Market Analysis evaluation of technology to be used, indicated that the gas turbine would be the generating unit used in practically all plants. In this regard, there has been a wider degree of discrepancy since, as seen above, while gas turbines are indeed used in the majority of cases, the contribution, particularly number-wise, of systems based on back-pressure steam turbines, is by no means negligible.

On a final note and by way of summarising this section, it can be said that on 40% of the time having elapsed of the total estimated necessary to convert into reality the potential for co-generation detected in Spanish industry, the degree of achievement is very close on said 40%; there are some small discrepancies, owing basically to the fact that average plant size and annual hours of operation are somewhat lower than foreseen.

6. CONCLUSIONS

Although from 1985 onwards there had been a greater contribution of co-generation to the sum total of electricity generated, the development of this technology up to 1987 was on a minor scale.

The level of co-generation in Spanish industry, expressed as the quotient between co-generated electricity and total electricity production, has, over the past two years, experienced growth estimated at 56%, going from 2.0% in 1987 to a figure of nearly 3.12% in 1989.

In spite of such a considerable increase, one must take into account that the average level of co-generation in 1985 in a 12-country strong EEC, was 8.13%; and therefore, standardisation with the EEC energy system still requires an additional effort if Spanish industry is to achieve a competitive structure and favourable economic growth.

The considerable development now taking place in this technology is being accompanied by a marked change in the parameters which set the profile for co-generating plants.

Thus new schemes mainly make use of gas turbines having an average size in the order of 5 MW and running for nearly 7 000 h.p.a. as their electricity generating unit.

The principal reasons which, in our judgement, serve to explain the present strong penetration of this alternative energy supply are as follows:

- The growing penetration of natural gas, which makes it possible for engine units having high electrical efficiency to be used.
- The margin between the cost of electricity and that of fuel, which decides whether or not the profitability of this kind of investment will prove attractive and which, until June of this year, had been following an unmistakably upward trend.
- The existence since 1982 of the Royal Decree for the Advance of Auto-generation (Real Decreto de Fomento de la Autogeneración) which regulates the conditions for transferring energy between auto-generators and the public grid.
- The support that has been forthcoming from the Government for these systems, by virtue of their being considered, from a national point of view, as an option meeting with the basic principles upon which an energy planning review can be based.

As there still exists a known potential for this technology in Spanish industry - put at a minimum of 350 MW to be installed by 1993 - the continuity of the present penetration of co-generating systems will doubtless be conditioned by the future development of the above factors.

KEY

KEY	SECTOR
3	PETROLEUM AND NATURAL GAS EXTRACTION, PETROLEUM REFINING
6	WATER CATCHMENT, PURIFICATION AND DISTRIBUTION EXCEPT IRRIGATION
8	MINERAL ORE AND ROCK MINING EXCEPT ENERGY RESOURCES
9	STEEL MANUFACTURING AND CASTING
12	BRICKS, ROOF TILES AND POTTERY
13	CHINA-WARE, PORCELAIN, REFRACTORY ARTICLES, FLOOR AND WALL TILES ETC
14	GLASS-MAKING INDUSTRIES
16	CHEMICALS INDUSTRY EXCEPT PETROCHEMICALS
21	AUTOMOBILE AND BICYCLE MANUFACTURING
23	FOOD, DRINK AND TOBACCO INDUSTRIES
24	TEXTILE AND CLOTHING INDUSTRIES
26	WOOD, CORK AND BULK TIMBER INDUSTRIES
27	PULP, PAPER AND CARDBOARD ,HANDLING & PROCESSING
28	GRAPHIC ARTS AND PUBLISHING
29	RUBBER PROCESSING
30	PLASTICS AND OTHER INDUSTRIES
35	GOVERNMENT ADMINISTRATION AND OTHER PUBLIC SERVICES

KEY	TYPE OF FUEL
F.O.	FUEL OIL
N.G.	NATURAL GAS
H.C.	HIGH GRADE COAL
B.L.	BLACK LIGNITE
OTHER	OTHER FUELS
O.P.P.	OTHER PETROLEUM PRODUCTS
REC.HEAT	RECOVERY OF RESIDUAL HEAT
L.C.	LOW GRADE COAL
B.F.G.	BLAST FURNACE GAS
O.GAS	OTHER GASES

KEY	SYSTEM
BPST	BACK PRESSURE STEAM TURBINE
DIESEL	DIESEL UNITS
CONDENS-ST	CONDENSED STEAM-TURBINE
GE-GT	GAS ENGINES OR TURBINES

TABLE O

ELECTRIC ENERGY PRODUCTION IN SPAIN

SOURCE:	1985 (GWh)	1986 (GWh)	1987 (GWh)	87/86 %
HYDROELECTRIC	33,033	27,415	28,167	+ 2.7
THERMOELECTRIC	66,286	64,277	63,952	- 0.5
NUCLEAR	28,044	37,458	41,271	+10.2
TOTAL	127,363	129,150	133,390	+ 3.3

TABLE 1

AUTOGENERATED ELECTRIC ENERGY PRODUCTION

SOURCE:	1985 (GWh)	1986 (GWh)	1987 (GWh)	87/86 %
HYDROELECTRIC	1,442	1,421	1,523	+ 7.2
CO-GENERATION	2,093	2,291	2,668	+16.5
TOTAL	3,535	3,712	4,191	+12.9

TABLE 2

LEVELS OF AUTO-GENERATION AND CO-GENERATION

	1985 %	1986 %	1987 %	87/86 %
SOURCE:				
HYDROELECTRIC	1.13	1.10	1.14	+ 3.6
CO-GENERATION	1.64	1.77	2.00	+13.0
TOTAL	2.77	2.87	3.14	+ 9.4

TABLE 3

CO-GENERATION IN THE AUTONOMOUS COMMUNITIES - 1987

AUTON. COM.	PRODUCTION (MWh)	PERCENTAGE s/TOTAL
ANDALUCIA	481,983	18.06
ASTURIAS	261,879	9.81
ARAGON	193,932	7.27
BALEARES	0	0.00
CANARIES	317,378	11.89
CANTABRIA	301,752	11.31
CASTILLA-LA MANCHA	277,187	10.39
CASTILLA-LEON	181,066	6.79
CATALONIA	254,029	9.52
EXTREMADURA	4,468	0.17
GALICIA	90,403	3.39
MADRID	0	0.00
MURCIA	58,858	2.21
NAVARRE	54,866	2.06
BASQUE COUNTRY	189,953	7.12
RIOJA	0	0.00
VALENCIANA	465	0.02
CEUTA-MELILLA	6	0.00
TOTAL	2,668,225	100.00

TABLE 4

DEGREE OF SELF-SUPPLY - 1987

AUTON. COM.	PRODUCTION (MWh)	CONSUMPTION (MWh)	SUPPLY %
ANDALUCIA	481,983	13,358,798	3.61
ASTURIAS	261,879	5,987,991	4.37
ARAGON	193,932	4,650,880	4.17
BALEARES	0	1,802,631	0.00
CANARIES	317,378	2,558,022	12.41
CANTABRIA	301,752	2,389,419	12.63
CASTILLA-LA MANCHA	277,187	4,425,511	6.26
CASTILLA-LEON	181,066	6,282,505	2.88
CATALONIA	254,029	21,901,631	1.16
EXTREMADURA	4,468	1,139,326	0.39
GALICIA	90,403	9,413,126	0.96
MADRID	0	11,596,866	0.00
MURCIA	58,858	2,468,544	2.38
NAVARRE	54,866	2,144,300	2.56
BASQUE COUNTRY	189,953	11,065,866	1.72
RIOJA	0	667,353	0.00
VALENCIANA	465	10,062,245	0.00
CEUTA-MELILLA	6	107,257	0.01
TOTAL	2,668,225	112,022,271	2.38

TABLE 5

TRENDS IN DEGREE OF SELF-SUPPLY

	1985 (GWh)	1986 (GWh)	1987 (GWh)	87/86 %
CO-GENERATED PRODUCTION	2,093	2,291	2,668	+16.5
NET CONSUMPTION	105,579	107,953	112,022	+ 3.8
SELF-SUPPLY (%)	1.98	2.12	2.38	+12.3

TABLE 6

CO-GENERATION SECTOR-BY-SECTOR - 1987

SECTOR	PRODUCTION (MWh)	PERCENTAGE s/TOTAL	CONSUMPTION (MWh)	SUPPLY (%)
3	528,667	19.81	1,543,476	34.25
6	209,453	7.85	1,486,529	14.09
8	5,942	0.22	1,386,312	0.43
9	353,863	13.26	9,403,087	3.76
16	337,128	12.63	8,097,659	4.16
23	221,763	8.31	4,596,879	4.82
24	84,815	3.18	2,921,226	2.90
27	837,774	31.40	3,286,958	25.49
35	88,820	3.33	3,631,115	2.45
TOTAL	2,668,225	100.00	36,353,241	7.34

TABLE 7

DISTRIBUTION OF CO-GENERATED PRODUCTION
ACCORDING TO SIZE

LEVEL OF PRODUCTION (MWh)	<3000	>3000 <5000	>5000 <10000	>10000 <30000	>30000 <100000	>100000	TOTAL
No. PLANTS (N)	5	6	15	15	17	7	65
PRODUCTION	4,278	26,427	101,199	260,813	1,022,211	1,253,297	2,668,225

TABLE 8

DISTRIBUTION OF PRODUCTION
ACCORDING TO TYPE OF FUEL

	H.C.	L.C.	B.L.	F.O.	O.P.P.	N.G.
GROSS PROD. (MWh)	200,263	155,655	412	1,220,808	51,067	296,703
%	7.51	5.83	0.02	45.75	1.91	11.12

	B.F.G.	OGAS	OTHER	REC.HEAT	TOTAL
GROSS PROD. (MWh)	107,407	72,602	400,754	162,554	2,668,225
%	4.03	2.72	15.02	6.09	100.00

TABLE 9

ACTIVE CO-GENERATED POWER CAPACITY

	POWER CAPACITY (KW)	HOURS
1985	707,400	2,959
1986	666,786	3,437
1987	753,262	3,542
87/86 (%)	+12.97	+3.05

TABLE 10

PLANT POWER CAPACITY IN THE AUTONOMOUS COMMUNITIES - 1987

AUTON. COM.	POWER CAPACITY (MWh)	PERCENTAGE s/TOTAL
ANDALUCIA	158,577	21.05
ASTURIAS	71,000	9.43
ARAGON	55,113	7.32
BALEARES	0	0.00
CANARIES	54,349	7.22
CANTABRIA	103,245	13.71
CASTILLA-LA MANCHA	54,000	7.17
CASTILLA-LEON	93,996	12.48
CATALONIA	69,681	9.25
EXTREMADURA	3,000	0.40
GALICIA	13,891	1.84
MADRID	0	0.00
MURCIA	16,145	2.14
NAVARRE	17,080	2.27
BASQUE COUNTRY	41,335	5.49
RIOJA	0	0.00
VALENCIANA	1,130	0.15
CEUTA-MELILLA	720	0.10
TOTAL	753,262	100.00

TABLE 11

PLANT POWER CAPACITY SECTOR BY SECTOR

SECTOR	POWER CAPACITY (KW)	PERCENTAGE s/ TOTAL
3	113,052	15.01
6	31,200	4.14
8	8,880	1.18
9	86,000	11.42
16	68,470	9.09
23	181,860	24.14
24	33,230	4.41
27	215,570	28.62
35	15,000	1.99
TOTAL	753,262	100

TABLE 12

DISTRIBUTION OF CO-GENERATING INSTALLATIONS ACCORDING TO SIZE

PLANT SIZE (MW)	<=1	>1 <=5	>5 <=10	>10 <=15	>15 <=25	>25
No.PLANTS (N)	5	16	23	9	7	5
POWER CAPACITY (KW)	2,972	53,623	172,762	115,251	153,909	254,745

TOTAL	
No.PLANTS (N)	65
POWER CAPACITY (KW)	753,262

TABLE 13

POWER CAPACITY BY REFERENCE TO TECHNOLOGY - 1987

SYSTEM	POWER CAPACITY (KW)	PERCENTAGE s/TOTAL	No. PLANTS
BPST	654,551	86.90	54
CONDENS.ST	39,280	5.21	5
DIESEL	8,950	1.19	3
GE-GT	50,481	6.70	3
TOTAL	753,262	100	65

TABLE 14

PLANTS IN OPERATION POST 1987

		NO.PLANTS (N)	POWER CAPACITY (MW)	PRODUCTION (MWh)
TOTAL		24	83.069	562,646
TECHNOLOGY	BPST	9	11.399	74,118
	GE-GT	15	71.670	488,528
AUTONOMOUS COMMUNITY	CASTILLA-LEON	3	8.600	55,797
	CATALONIA	10	47.970	323,406
	GALICIA	1	0.324	867
	MADRID	1	1.000	7,564
	BASQUE COUNTRY	4	11.205	73,875
	VALENCIANA	5	13.970	101,137
SECTOR	3	1	1.500	9,330
	12	1	0.400	2,564
	13	1	3.700	31,344
	14	1	1.200	9,500
	16	7	37.489	243,959
	23	2	4.270	35,514
	24	2	1.470	5,420
	26	1	3.400	16,078
	27	7	28.640	201,242
	28	1	1.000	7,695
FUEL	H.C.	2	1.470	5,420
	F.O.	1	1.500	9,330
	N.G.	17	75.910	519,340
	REC.HEAT	4	4.189	28,556

TABLE 15

PLANTS UNDER CONSTRUCTION

		NO.PLANTS (N)	POWER CAPACITY (MW)	PRODUCTION (MWh)
TOTAL		18	125.970	927,568
TECHNOLOGY	BPST	1	1.000	6,500
	GE-GT	17	124.670	921,068
AUTONOMOUS COMMUNITY	ANDALUCIA	1	51.000	389,760
	ASTURIAS	1	1.000	8,647
	ARAGON	1	9.000	68,700
	CASTILLA-LEON	1	0.490	3,533
	CATALONIA	5	18.820	132,909
	MADRID	1	5.000	40,000
	BASQUE COUNTRY	3	27.400	193,892
	VALENCIANA	5	13.260	90,127
SECTOR	3	1	51.000	389,760
	12	1	0.490	3,533
	13	3	8.640	53,942
	16	4	24.920	189,991
	21	1	3.620	29,685
	27	6	27.700	202,904
	29	1	8.600	50,892
	30	1	1.000	6,861
FUEL	H.C.	1	1.000	6,500
	N.G.	16	73.970	531,308
	OGAS	1	51.000	389,760

TABLE 16

STATE OF CO-GENERATION - 1989

		NO.PLANTS (N)	POWER CAPACITY (MW)	PRODUCTION (MWh)
TOTAL		107	962.301	4,158,439
TECHNOLOGY	BPST	64	666.950	2,353,969
	CONDENS.ST	5	39.280	207,082
	DIESEL	3	8.950	8,281
	GE-GT	35	247.121	1,589,107
AUTONOMOUS COMMUNITY	ANDALUCIA	14	209.577	871,743
	ASTURIAS	3	72.000	270,526
	ARAGON	3	64.113	262,632
	BALEARES	0	0.000	0
	CANARIES	3	54.349	317,378
	CANTABRIA	4	103.245	301,752
	CASTILLA-LA MANCHA	2	54.000	277,187
	CASTILLA-LEON	21	103.086	240,396
	CATALONIA	22	136.471	710,344
	EXTREMADURA	1	3.000	4,468
	GALICIA	2	14.215	91,270
	MADRID	2	6.000	47,564
	MURCIA	2	16.145	58,858
	NAVARRE	2	17.080	54,866
	BASQUE COUNTRY	14	78.940	457,720
	RIOJA	0	0.000	0
	VALENCIANA	11	28.360	191,729
	CEUTA-MELILLA	1	0.720	6
SECTOR	3	10	165.552	927,757
	6	2	31.200	209,453
	8	1	8.880	5,942
	9	3	86.000	353,863
	12	2	0.890	6,097
	13	4	12.340	85,286
	14	1	1.200	9,500
	16	16	130.879	771,078
	21	1	3.620	29,685
	23	29	186.130	257,277
	24	4	34.700	90,235
	26	1	3.400	16,078
	27	29	271.910	1,241,920
	28	1	1.000	7,695
	29	1	8.600	50,892
	30	1	1.000	6,861
	35	1	15.000	88,820
FUEL	H.C.			212,183
	L.C.			155,655
	B.L.			412
	F.O.			1,230,138
	O.P.P.			51,067
	N.G.			1,347,351
	B.F.G.			107,407
	OGAS			462,362
	OTHER			400,754
	REC.HEAT			191,110

TABLE 17

EFFECTIVE POTENTIAL: AIMS AND ACHIEVEMENTS

	No. PLANTS (N)	ACHIEV.	POWER CAPAC. (MW)	ACHIEV.	PRODUCT. (GWh)	ACHIEV.
TOTAL	102	41%	585	36%	4,321	34%
ANDALUCIA	8	13%	65	78%	469	83%
ASTURIAS	2	50%	30	3%	233	4%
ARAGON	4	25%	20	45%	157	44%
CANTABRIA	4	0%	66	0%	498	0%
CASTILLA-LA MANCHA	1	0%	21	0%	173	0%
CASTILLA-LEON	6	67%	30	30%	215	28%
CATALONIA	33	45%	158	42%	1,185	39%
GALICIA	1	100%	12	3%	94	1%
MADRID	5	40%	23	26%	169	28%
NAVARRE	2	0%	6	0%	39	0%
BASQUE COUNTRY	22	32%	132	29%	926	29%
RIOJA	1	0%	1	0%	4	0%
VALENCIANA	13	77%	21	130%	159	120%
3	7	29%	82	64%	650	61%
9	1	0%	50	0%	342	0%
11	3	0%	3	0%	28	0%
13	5	80%	6	206%	44	194%
14	2	50%	5	24%	41	23%
16	24	46%	142	44%	1081	40%
17	3	0%	49	0%	346	0%
21	1	100%	2	181%	12	247%
23	14	14%	53	8%	382	9%
24+25	8	25%	20	7%	120	5%
26	1	100%	3	113%	20	80%
27+28	28	50%	134	43%	1022	40%
29	4	25%	34	25%	221	23%
OT	1	300%	2	95%	12	108%

TABLE 18

OVERVIEW OF TECHNOLOGIES

DISCUSSION

SUMMARY

PARTICIPANTS

The following participants have asked questions or made comments:

GREEN, D., Combined Heat and Power Ass. (U.K); FERNANDEZ ZORRILLA, A., Iberduero (Spain); GREY,R., Building Services Research and Information Ass. (U.K.); ALBISU, F., Sener S.A. (Spain); KORRES,C.J., (C.E.C.); HODES, ESYS (France); SAYANS, F., Deutz MWM S.A. (Spain); BERKELMANS,F., Royal Schelde (The Netherlands) and MARANIELLO,G., Ansaldo (Italy).

SPEAKERS

Answers were given by:

ALBISU,F., Sener S.A. (Spain): CONTRERAS,D., IDAE (Spain) and GYFTOPOULOS, E., M.I.T. (U.S.A.).

TOPICS DISCUSSED

- Wood and diesel fueled cogeneration costs in the Phillipines and U.S.A.
- Reciprocating engines' efficiency.
- Relative competitiviness between wood, diesel and gas fired cogeneration systems in the U.S.A.
- Overview of the Government solid waste Cogeneration Policy in Spain. National, Regional and Municipal duties, ownership of the operator's companies and project developers.
- Marine applications of cogeneration in the U.S.A. and Spain.
- Costs of average and small size cogeneration systems in Spain.
- Environmental and private cost reductions due to cogeneration.
- Incremental and total fuel consumption in cogeneration electricity production vs public utilities production.
- Outlook of cogeneration power capacity in Spain.
- Government support for cogeneration in Spain.

COMMENT

There was great interest about the investment and associated costs figures, the outlook and support of government's for the different cogeneration systems in the different countries. No single figure could be given because it depends on the availability and price of fuels in each country, culture of the society, energy demand profile, government's policies and so on. What was very clear was that, regardless of countries' characteristics, cogeneration has a great future, is very profitable and that these countries peculiarities determine the opportunity and type of system to be used.

COGENERATION FINANCING AND LEGISLATION IN E.E.C. AND THIRD COUNTRIES

Chairman : Mr.I. Carrera Ezpeleta

Industry and Transport Director

I D A E

THE HISTORY AND STATUS OF FINANCING COGENERATION PROJECTS IN CALIFORNIA WITH PROSPECTS FOR THE FUTURE

JAN HAMRIN, PhD

Independent Energy Producers Association
Jan Hamrin Associates
Mill Valley, Ca.
U.S.A.

SUMMARY

This paper chronicles the history of cogeneration development in the State of California, USA, and outlines the conditions necessary for its development. Of particular interest is the financing of cogeneration projects in the United States including cashflow analysis, risk analysis and mitigation. The types of financing structures most commonly used are described along with their risk allocation characteristics. Finally, data is presented for cogeneration projects currently on-line in California and Texas which shows their reliability exceeds that of the average utility project. However, since concerns about future project reliability remain, the paper examines methods which can be used to reduce these concerns and to reduce the probability of projects being cancelled.

RESUMEN

Esta ponencia expone la historia del desarrollo de la cogeneración en el Estado de California, E.E.U.U., y pone de relieve las condiciones necesarias para su desarrollo. Particularmente interesante es la financiación de los proyectos de cogeneración en los Estados Unidos incluyendo análisis de cash-flow, análisis de riesgos y disminución de los mismos. Se describen los tipos de estructuras de financiación más frecuentes así como su distribución de riesgos. Finalmente se presentan los datos relativos a proyectos actualmente operativos en California y Texas y que demuestran que su fiabilidad es superior a la media de los proyectos de las compañías eléctricas. Sin embargo, dado que existen incertidumbres sobre la viabilidad de futuros proyectos, la ponencia analiza métodos para reducir estas incertidumbres y la probabilidad de que los proyectos se cancelen.

THE HISTORY AND STATUS OF FINANCING COGENERATION PROJECTS IN CALIFORNIA WITH PROSPECTS FOR THE FUTURE

Jan Hamrin, PhD

Executive Director, Independent Energy Producers Association and President, Jan Hamrin Associates

P. O. Box 40
Mill Valley, CA., U.S.A. 94920

1. HISTORIC BACKGROUND

In 1978 when the Public Utility Regulatory Policy Act (PURPA) was passed there was very little cogeneration in existence in the United States and virtually no industry available to build any. By 1989 the industry has mushroomed into a multi-billion dollar business. While projects had difficulty finding financing in 1978, good projects have their pick of investors in 1989. This amazing growth of the independent power industry in the United States is due to several factors:

* Passage of the Public Utility Regulatory Policy Act (PURPA) in 1978 which created the legal foundation for regulatory action
* Entrepreneurs willing to take on significant risks to develop the early projects
* Increasing energy prices which focused attention on alternatives to conventional power generation
* The oil crisis of the late 1970's and continuing problems of developing nuclear power
* Environmental concerns which emphasized the more efficient use of energy

Conventional technologies such as gas-fired cogeneration, grew the most rapidly because of the thousands of opportunities for its application, the equipment was commercially proven and available, and the risks were perceived as low. Other technologies such as wind, solar and biomass had little or no technical track-record and were encouraged through tax-credits which provided more opportunities for financing and allowed technical research and development to take place in the field as projects were built and gained operating experience.

California, because of its size, energy resources and political climate took an early lead in the development of renewable energy technologies and cogeneration. Since 1978, 7,344 MW of cogeneration, biomass, geothermal, small hydro, solar electric and wind generation technologies have come on line in California and another 3-5,000 MW are scheduled to come on line in the next two years. Of this amount, 4,720 MW are from Cogeneration/Biomass projects. Cogeneration and renewable energy technologies represent three times as much power as the recently completed Diablo Canyon Nuclear Generation Units (2200 MW), and were constructed in one third the time at approximately one-half the cost.

Texas with its large petro-chemical industries, a long history of the use of cogeneration and its policy of requiring transmission access for wholesale power transactions leads the nation in the development of cogeneration facilities with 7473 MW of independent generation on line.

Federal and State tax credits were a major factor in the financing of many of the early renewable energy projects. However, energy tax credits were **not** offered for fossil fueled cogeneration projects which has been one of the most vigorously developed of the PURPA technologies. This paper outlines the conditions necessary for the development of independent power generation projects; explains how cogeneration projects are financed in the United States, presents data on the reliability of such projects to the present time, and projections for future development.

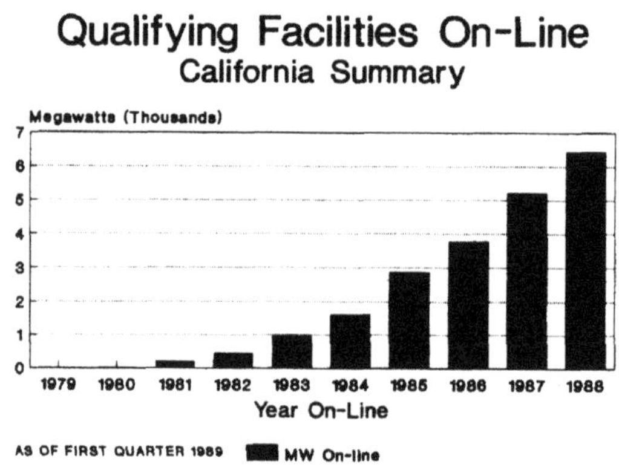

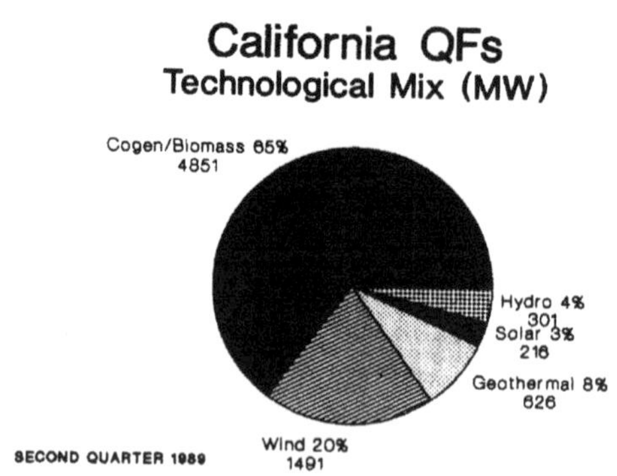

2. CONDITIONS NECESSARY FOR THE DEVELOPMENT OF INDEPENDENT POWER GENERATION

There are basically two different types of generation facilities: 1) A "stand alone" or self-generation facility which generates electricity entirely for its own internal use; and, 2) A generation facility which sells excess electricity to other users or into the utility grid. In the first case all that is needed is the legal right to install such a facility and sufficient energy savings to make the investment in generation cost effective for the company involved.

The second case is much more complicated. Let us assume the generation facility wishes to sell excess electricity into the utility grid. In addition to laws permitting this type of action, several other conditions are necessary before a project can be financed and built. First there are regulatory/contractual needs:

* **Clear and Equitable Interconnection Specifications.** Though the utility usually specifies the design and equipment required to interconnect the independent producer with the utility system, such specifications should be clear, fair and consistent with common utility practice. Such interconnection facilities should not be "over engineered" to artificially inflate the cost to the independent producer.

* **A Stable Fuel Supply Over the Life of the Project.** It is critical that a cogeneration project have a stable fuel supply. Where some or all of that fuel is purchased from the utility (or some other government agency), the escalation rate for that fuel price should be no more than the escalation of the buy-back rate for the power. Fuel supply contracts can assure producers and utilities of reliable feedstocks.

* **Backup Power.** If the generation facility is designed primarily for self-generation then it will probably include some sort of backup generator. However, a cogeneration facility may depend upon the local utility for backup or standby power when the plant is out for scheduled maintenance or for a forced outage. Any standby rates or demand charges should be based upon the cost of providing such service. If standby rates and demand charges are set too high, the cogeneration facility may either cease generating or provide its own standby facilities and leave the utility system all together.

* **Stable Regulatory Environment.** Sanctity of the power purchase contract and other contracts is critical. There should be no danger that a contract will be changed or cancelled due to a change in political or regulatory players. If developers are afraid that the "game" may be changed due to an unstable regulatory environment, they will not participate.

* **A Financiable Contract.** A "financiable contract" is one which includes a predictable and sufficient revenue stream, clear and equitable interconnection specifications, unbiased standby rates and demand charges, and does not include open-ended liabilities or assign risks to the project over which the developer has no control. A standard power purchase agreement containing the basic terms and conditions which can serve as a basis for negotiation of special provisions is particularly important in encouraging development.

All of these elements are critical for the successful development of a private power project. Eliminate any one and it will be very difficult to obtain private sector participation or financing. Private power developers are business people operating in a complex business environment. Financiable projects are those that can demonstrate returns commensurate with risks.

3. FINANCING COGENERATION PROJECTS IN THE UNITED STATES

Assuming all the conditions listed above are present (including a sufficient price for the electricity to be sold), there is yet another set of hurdles to be overcome for a project to be financed. Anyone financing a project deals in risks. A project financier must either minimize the probability of a particular risk situation occurring or hand the risk off to someone else. In order to secure financing, a financier must be assured that the project will be:

* Completed on time, within budget and to specification

* Operate successfully

* Generate sufficient cash to repay the financing

In evaluating a project, potential financiers will focus on three major areas:

* Track records of the principal parties

* Project cashflow projections and economic analysis

* **Project risk assessment**

The first two areas are fairly straight forward. <u>**Track Record:**</u> the financier must feel confident that the project's design and construction will meet standards, that the operator will be successful and that the company will successfully bring together all other requirements of the project. The financier will also examine the financial condition of the major parties. As discussed later, a primary means of risk mitigation is to transfer the risks contractually through fuel, construction, operations and maintenance contracts. This risk transfer is worthless unless the parties are economically and technically capable of assuming and mitigating such risks.

<u>**Cashflow Analysis:**</u> several measures are used to evaluate a project's expected financial performance including discounted cash flow analysis, payback periods and internal rates of return. The lender or lessor will focus particularly on debt coverage ratios. This is the amount of cash available annually after payment of all fuel, operating and overhead expenses but before payment of the project debt. Generally a debt coverage ratio of at least 1.3 is required for a project to be financed.

<u>**Risk Analysis and Mitigation:**</u> project financiers attempt to foresee every possible way that something could go wrong and then divide, allocate and mitigate those risks. For example:

* **Completion Risk** - These are mitigated through engineering, procurement and construction contracts which frequently contain early completion bonuses; late completion damages; independent technical, budget and schedule evaluation; a performance bond for the entire contract; funding increments tied to milestone achievements; and stringent acceptance testing provisions.

* **Performance and Operating Risks** - These are mitigated through a long term fuel supply plan and contracts and through operation and maintenance contracts, equipment warranties and guarantees; overhaul and spare parts capital set asides, and insurance against natural disaster.

* **Market and Pricing Risk** - This is mitigated through long term power purchase agreements with energy output prices indexed to inputs, fixed and variable revenues matching respective costs and a contractual obligation for the utility to buy output. The thermal energy contract is important as well. The project financier is concerned both with the credit-worthiness of the thermal

purchaser, the strength of the thermal sales contract and alternative markets for the output.

* **Regulatory and Environmental Risks** - These risks are mitigated through 1) High level governmental decrees; 2) Institutionalization of private power regulations; 3) Statement of commitment by utilities; and, 4) One source information on environmental permits.

Financing Structures: The most conventional way of financing a project is through some combination of debt and equity. Debt:equity ratios can vary widely but are normally between 50:50 and 90:10. The lender usually has additional security through recourse to the project sponsors assets above and beyond those associated with the project. The lender can appropriate those assets if the project fails. Non-recourse project financing (in which the investor provides 100% of the project's financing with recourse only to the project and its assets) is an alternative to this. There are several types of financiers. The most common types are:

* **Full Recourse Lenders** - They may finance both equipment and construction costs and may be willing to provide 80% or more of the projects costs.

* **Equipment Financiers** - They only finance the equipment (which is generally less then 50% of the cost of the project). The purchaser must normally guarantee repayment whether a project fails or succeeds.

* **Equity Investors** - Generally provide financing beyond that available from equipment financiers or conventional lenders. Equity investors bear the ultimate project risk. A project's developers/sponsors often provide most of the equity investment.

* **Non-recourse Lenders** - Have recourse only to the assets invested in that particular project. This requires a strong project and frequently greater equity investments by the project's sponsors.

* **Project Finance Lenders** - "Project financing" is one in which the investor provides 100% of the project's financing with recourse only to the project and its assets. The key to mitigating the project financier's risks is secured through various contractual arrangements as discussed above.

Leasing is also a common method of financing a project. Leasing can have as many variations as the other types of financing listed above. May cover only the equipment or up to the entire project cost.

Negotiating the financing package and the contracts which support it are the most critical roles of the project development team. No one type of financing mechanism dominates the United States independent generation projects developed over the last decade.

4. PROJECT RELIABILITY

Reliability refers to the probability of a resource being available when needed to serve load. This topic can be divided into two parts: 1) The probability of a project coming on line when needed; and, 2) the consistent availability of the project once it has come on line.

The Probability of a Project Coming on Line When Needed: Whether a new generating project ever becomes operational depends upon a number of things such as a) the ability of the project to get financing, b) receive siting and permit approvals, and c) be constructed within the necessary timeframe. Since independent energy projects in the United States are not paid until they are on-line and generating, the concern about a particular project coming on line is a planning one. Three approaches can be taken to manage this risk: i) include contract language which makes the project liable for any financial damages incurred due to a projects lateness or failure to come on line; ii) trying to predict and screen applications to find those projects most likely to be successful and iii) planning for a specified attrition rate in the expectation that some projects will fail. All three approaches have been used in the U.S.; the first (accountability for damages) and the third (planning for attrition) have been the easiest to implement and the most successful.

The issue of projects not coming on line is possibly of greater concern to other independent energy developers than to utilities or government in that "phantom" projects may prevent those with viable projects from being allowed to develop.

Reliability of On-line Projects: Two types of data can be used to evaluate the reliability of on-line projects: aggregate data on projects which have come on line since PURPA was passed and anecdotal data on the few cogeneration projects which predated PURPA and have been on line for several decades.

Because PURPA was passed just ten years ago, many states are only now seeing projects come on line. However, California (with 7,344 MW from 989 projects) and Texas (with 7473 MW of on-line generation) do have sufficient projects and experience to provide valuable data.

California utilities offer to purchase power under firm capacity and as-available capacity contracts. According to data filed by PG&E in its 1988 electricity costs reasonableness proceeding, the 1621 MW of firm capacity projects (primarily cogeneration, biomass and geothermal projects) on line at that time were operating at a 94.8% capacity factor. (1)

In Texas, results for cogeneration projects were obtained from a survey conducted by the Gulf Coast Cogeneration Association in 1987 which indicated availability and capacity factors of 96% and 84%, for the 3126 MW of capacity surveyed. (2) The Gulf Coast Cogeneration survey includes systems which have been in service since 1929. Collectively, they have maintained an 88% capacity factor. Survey results indicate "cogeneration systems continued to operate through cycles of business conditions and fuel pricing changes during the 1970s and 1980s and are operating today, sometimes under much different conditions than originally envisioned by the project initiators." (3)

Another area of interest is delivery of power during times of emergency. One illuminating example is the 1988 experience with Hurricane Gilbert in south Texas, particularly in the Houston Lighting & Power service territory. Information available (4) now indicates cogenerated power in HLP's service area continued without interruption during the storm threat and in some cases cogenerators actually increased their net energy flows to the utility. Further, gas supplies to Texas cogenerators appeared to be more reliable than HLP's offshore natural gas suppliers, which were forced to interrupt deliveries, causing shutdowns of some of HLP's thermal generation. In some cases cogenerators were unable to deliver electricity due to the loss of the utility's transmission lines, but overall the independent cogenerators provided critical backup electricity during this emergency. Similar responses from third party producers were observed during an outage of a major northern California intertie line in the spring of 1984 and in Southern California during power shortages in February, 1989. Independent electric generators helped meet the needs of 1.8 million California customers during the Southern California Edison 1989 emergency. (5)

A critical factor in the ability of independent generators to continue supplying power during emergencies appears to be the engineering of the interconnection facilities and confidence that neither equipment nor personnel will be endangered by remaining on the grid. However, it is precisely because of the disaggregated nature and small size of third party generation that it can play an important support role during unscheduled outages. In fact, many facilities can generate energy beyond their contracted limits during emergency situations if the utility is willing and able to take the power. The key is good communications between independent generators and the utility.

Historically, on-line independent generation plants have an excellent record of reliability, surpassing that of conventional utility plants, as indicated above. Yet there is concern among some regulators and utility managers that independent generation projects may not continue to operate at these levels in the future because of economic or resource risks. The following events are the ones most often discussed:

- major changes in ownership
- general failure of the economy or a specific industry
- disruption of fuel supply
- generic design flaw or
- contract design inappropriate to actual events

Fortunately, contract design and mixed portfolios of contract and project types can be used to resolve or hedge against these risks. (6) For example: Major changes in ownership need not affect a project's ability to perform. If project revenue exceeds operation and maintenance costs, someone will continue to operate the project. One of the considerable benefits to ratepayers of contracting for nonutility supply is the economic benefits of "pay for performance contracts." With independent electricity suppliers, ratepayers only pay for what they get. They are relieved of the risks of cost overruns, plants that do not operate as planned, and uncontrolled repair and replacement costs. Though some project owners and investors may loose money, the utility and the ratepayer should not if contracts are carefully crafted and utilities invest in a mixed portfolio of contract resources. Governmental institutions and utilities have tremendous flexibility in the design of programs and contracts for power which can provide the ratepayer economic benefits while hedging against the risks of economic uncertainties.

5. FUTURE PROJECTIONS

Independent power generation projects, especially cogeneration projects continue to be built in the United States. The benefits of "pay for performance" contracts with non-utility generators, fuel efficiency and the use of waste products as fuels (which would otherwise be a liability to be disposed of), the economic benefits to businesses which install cogeneration and the thousands of industrial and commercial applications appropriate for cogeneration makes this particular technology a continuing option for the future. In addition, environmental concerns such as acid rain and potential climate change effects from increasing amounts of greenhouse gases, make cogeneration an attractive transitional technology along with renewable energy generation for the electricity needed after other energy saving measures have been applied.

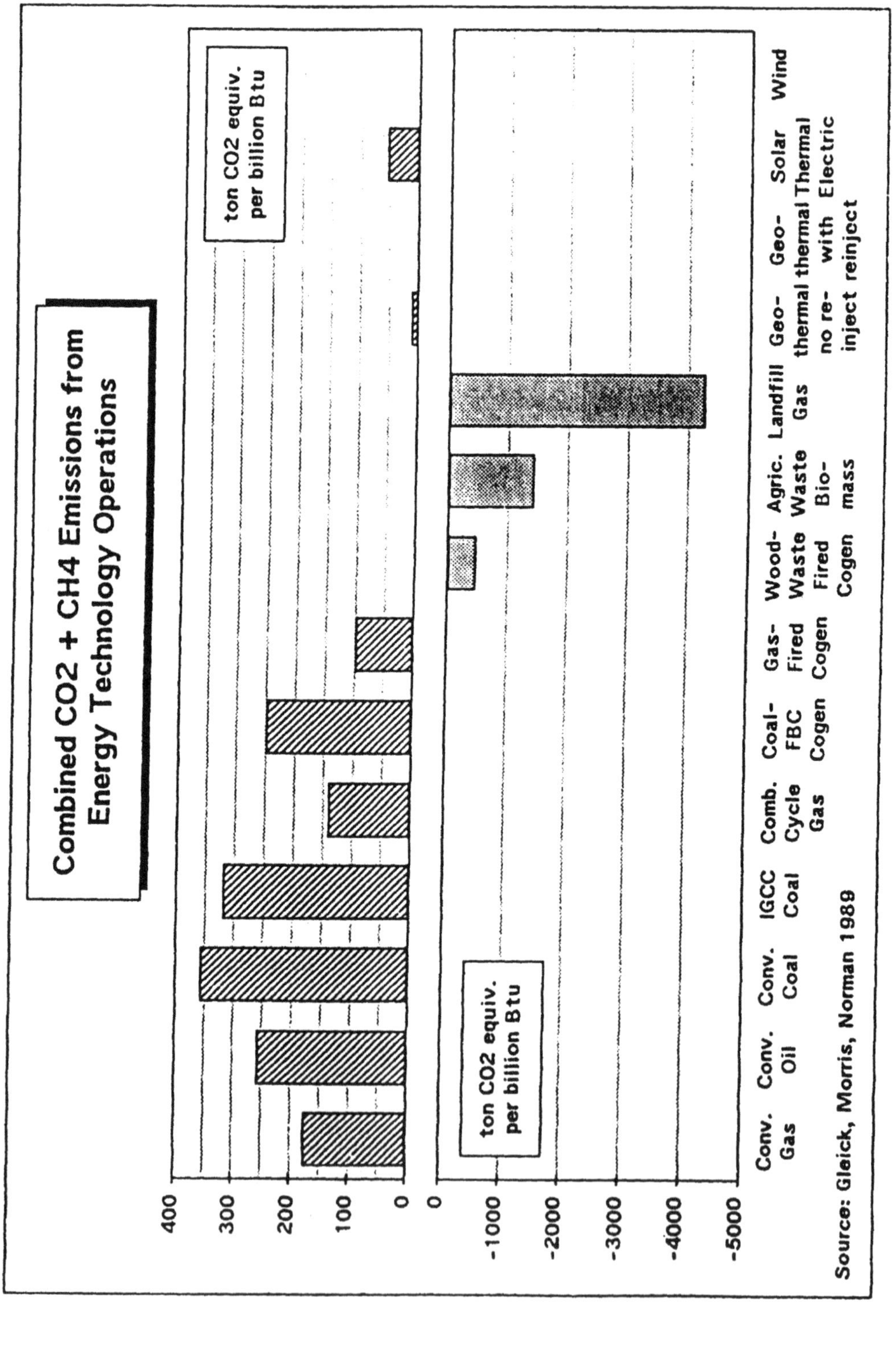

Source: Gleick, Morris, Norman 1989

The debate in California and elsewhere in the United States today is not whether to use cogeneration and other types of independent generation technologies but rather how best to structure contracts and prices and how to quantify and integrate environmental benefits into the pricing calculations to improve the efficient acquisition of these resources.

REFERENCES:

(1) This can be compared with large utility coal plants which average 75-80% availability, large nuclear plants (over 600 MW) which average 55-60 % availability, and utility gas or oil fired thermal plants which may run at 80-90% availability factors. As utility plants get larger, their reliability tends to go down.

(2) GULF COAST COGEN. ASN., SURVEY OF COGENERATION IN TEXAS 1 (1987).

(3) Id., at 1.

(4) Gulf Coast Cogen'n Assn. Newsletter (Oct. - Nov. 1988); see also Texas Industrial Electric Consumers (TIEC) Special Report (October 13, 1988); letter from J.E. Brunt (Dow Chemical) to L.G. Brackeen (Houston Lighting & Power) (Oct. 4, 1988).

(5) Letter from Southern California Edison Company, Feb. 9, 1989.

(6) For a longer discussion of reliability of nonutility power see "Nonutility Power and the Reliability Issue," by Jan Hamrin. THE ELECTRICITY JOURNAL, June, 1989 Volume 2, Number 5.

THIRD PARTY FINANCING

Dr. DEREK A. FEE

Directorate General for Energy
Commission of the European Communities

SUMMARY

The achievement of the Council's 1995 energy efficency objectives will require investments on a level far greater than that which is currently taking place. The potential market within the Community for third party financing services has been estimated at 86 billion ECU. The size of this potential market should act as sufficient incentive for the creation of ESCOs but this has so far not been the case. The paper describes several Novel Financing Mechanisms, the barriers to innovative energy efficiency financing in the European Community, what actions the Commission has taken to promote the use of third party financing and the special merits of third party financing for cogeneration projects.

RESUMEN

La consecución de los objetivos de eficiencia energética, declarados por el Consejo para 1995, requieren unas inversiones superiores a las actuales. El mercado potencial en la Comunidad Europea se estima en 86.000 millones de ECUS. El tamaño de este mercado potencial debería ser un incentivo suficiente para la creación de Sociedades de Servicios Energéticos (ESCO's) aunque esto no se haya producido hasta el momento. La ponencia describe varios mecanismos nuevos de financiación, las dificultades con que se encuentra la financiación innovadora en eficiencia energética en la Comunidad Europea, qué acciones ha tomado la Comunidad para impulsar la financiación por terceros y la idoneidad de la financiación por terceros para los proyectos de cogeneración.

THIRD PARTY FINANCING

by Dr. Derek A. Fee

Directorate-General for Energy

Commission of the European Communities

1. Introduction

The Council of Ministers, at their meeting in September of 1986, set new energy objectives for 1995, which included a further improvement in energy efficiency of at least 20%. The achievement of this improvement will be effected both by managerial and behavioural changes, and by investments.

Managerial and behavioural changes fall into two categories. The first category is better maintenance and control e.g. periodic cleaning and surveillance, improved fault detection, and better production planning. The second category is the changes in energy services, e.g. lowering thermo-stats, car pooling, and less hot water consumption.

Integrated energy efficiency investments are directed primarily towards purposes other than the rational use of energy e.g. new electrical appliances, new cars, new buildings, new burners and boilers, and new industrial processes. In these cases energy efficiency is only one of the factors being considered. These types of efficiency improvements are least likely to be affected by short-term energy prices or economic changes.

Discrete conservation investments are primarily or solely directed towards improving the end-use efficiency, can be expected to be most affected by short-term energy prices. If the price decline threatens the anticipated economic viability of the investment, it is likely to be postponed or possibly rejected.

Governments can best influence managerial and behavioural changes by carefully organised campaigns aimed at disseminating energy efficiency information or raising awareness about wasteful energy practices. Energy efficiency investments can be influenced by the provision of R&D grants aimed at spurring technological innovations which can make a significant contribution to energy efficiency, and by the provision of grants, fiscal incentives and soft loans for carrying out discrete energy efficiency investments. Another driving force for all energy efficiency improvements is, of course, the actual price paid for energy. The Community and the Member States have instituted a series of energy saving programmes of both an informational and investment incentive nature. These programmes were successful in improving the rational use of energy in Europe by 20% during the period 1974-85.

A recent study[1] carried out by the Commission has estimated that economically achievable energy efficiency investments, i.e. rate of return of 30%, represented a total European Community market of 86 billion ECU. This is made up of 44 billion ECU in the industrial sector and 42 billion ECU in the building sector.

One may assume that discrete energy efficiency investments make up only a part of the total investment required, and that measures in managerial and behavioural change, and integrated investment will continue to bear fruit. Nevertheless, the sheer scale of the required investment necessitates the development of financial instruments, other than direct State intervention, which will assist in accelerating the discrete investment in energy efficiency.

A Commission communication entitled ' Towards a European Policy for Energy Efficiency in the Industrial Sector'[2] has already examined some of the factors which militate against discrete energy efficiency investments, these include:

- low energy prices;

- the low priority often attached to energy saving investments in decision making processes;

- lack of knowledge of consumption;

- financial structure of firms, lack of finance; and

- the disparity of required rate of return between energy supply and energy savings projects.

For a novel financial mechanism to be successful it must counter all , or most of these factors.

2. Novel Financing Mechanisms.

Several financial mechanisms have been developed to accelerate energy efficiency investments. These include:

- innovative vendor financing, e.g. financial savings guarantees, vendor backed equipment leasing, package financing, and shared saving contracts;

- energy service company financing, e.g. third party financing;

- energy project financing; and

- utility financing.

Each type of financing uses different mechanisms, involves various technologies, and can involve more than two participants at the contractual level.

Vendor backed equipment leasing - where the lessor is the vendor himself- provides two key features of innovative financing. Firstly it does not require the lessee to provide the investment capital, so that it does not have to be included in the balance sheet and secondly it allows the buyer the opportunity of including some "caveat" conditions in the lease contract. This latter point may be important in limiting some of the risks which a purchaser would be exposed to. This technique represents an adaptation of conventional leasing to the energy efficiency sector and requires no further explanation.

On the other hand **energy service company financing, or third party financing,** is a new technique and has the potential for providing, at low risk, capital to enable discrete energy efficiency investments to be made. This mobilisation of private capital is accomplished by the operation of an **energy service company(ESCO)** obtaining the finance to fund an energy saving programme using the cost savings themselves to service the capital and to pay for that investment. Therefore, the energy savings are viewed as a ' stream of income' which can support a business: the business of investing in, and providing performance guarantees for energy conservation, by the ESCO.

The concept of the energy service company is, of course, central to the successful operation of the third party financing mechanism. An ESCO must provide a combination of engineering, financial and marketing skills. It must be capable of carrying out detailed energy audits, and of selecting technologies which would be suitable for achieving remunerative energy savings. Project finance must be raised, and the flow of funds from the project should be sufficient to repay the provider of the finance, and ensure the profitable operation of the ESCO. In general an ESCO has been defined as a company which ' **provides the service of auditing, installation, operations, maintenance and financing on a turnkey basis**. A company which sells equipment but which does not finance or maintain that equipment does not correspond to the definition of an energy service company.

The necessary steps to establish a third party financing investment are as follows. The ESCO carries out a rapid initial "walk through" energy audit to establish the likely level of possible energy savings. An outline proposal is then made to the facility owner which sketches out a programme for accomplishing these energy savings. A contract is negotiated, and a energy baseline or average consumption pattern is developed. The ESCO then carries out a detailed energy audit, and then installs equipment aimed at accomplishing the identified potential energy savings. The facility owner and the ESCO share the financial benefit from energy savings made during the term of the contract. Provision is normally made for adjustments to be made to the terms of the contract any time during the life of the contract. When the contract expires, the facility may renew the contract at an adjusted share of savings, he may become the outright owner of the equipment, or in some cases may have an option to purchase the equipment at a price decreed by the contract.

Third party financing **or "shared saving contracts"**(as they are often called in the United States) were initially conceived in North America where they were introduced in 1981. The market for third party financing in the United States has been developed over the period 1981-1986. In 1980 there were about 20 companies offering 'energy services' in the United States. Energy saving investments made through these companies resulted in about $1m being invested. By 1984, the number of companies had grown to 150, and annual investment stood at some $350m. By 1986 there were over 250 service companies offering to fund third party financing investments.

One of the factors which has assisted the growth of the 'energy services' market in the United States has been the active role played by government-Federal, State, and local. The active participation of government institutions has led to a situation where by 1985 the public sector accounted for 50% of all third party financing compared to 20% in 1983. At U.S. Federal level, the government has, through it's various departments promoted the use of third party financing in making energy saving investments in government buildings. The Federal Energy Management Programme has set up a clearing house on third party financing, in order to assist government building managers to avail themselves of the technique.

At State level, programmes have been developed to guide building managers on the utilisation of third party financing to reduce energy consumption in State run buildings.

At local level, many County administrations have supported schemes aimed at demonstrating the efficacy of third party finance for energy saving investments in public buildings and in individual homes.

Since the inception of the third party financing technique in the United States, many different organisations have entered the field to provide third party financing services. They include; engineering consultants, equipment manufacturers, subsidiaries of gas and electric utilities and, in some cases, local government itself.

In Europe, the concept of third party financing has been much slower to develop. A study carried out for the Commission in 1986 found that the technique is very little practiced . In 1985, the two large and several small European ESCOs collectively invested about 16 million ECU in energy saving projects. Only four countries, the United Kingdom, Belgium, Spain and Luxembourg had any direct involvement in third party financing while France and Italy had experience with financing techniques having some similar features. The 1985 investment figure can be contrasted with our estimated EUR-12 potential market of 86 billion ECU.

Third party financing has the following main advantages;

- the facility owner does not have to raise capital to finance conservation measures;

- the third party assumes all the risk that energy savings will occur;

- the facility owner does not have to determine which equipment is most appropriate for their facility;

- the facility owner can still make other investments while reaping the benefits of energy saving;

- it is usual for the facility owner to own the equipment at the end of the contract or arrangements can be made to secure equipment ownership.

The disadvantages of third party financing are include:

- third party financing contracts **tend to be complex**, resulting in a number of facility owners being discouraged from attempting such schemes;

- the **lack of historical energy consumption data** can become a limiting factor in the conclusion of a contract; and

- the economics of third party financing are often only justifiable **for large programmes** i.e. where the investment exceeds 100000 ECU.

<u>Energy project financing</u> is also a North American concept which has thusfar not been applied in Europe. The concept involves the construction of an energy producing plant by a partnership which typically invests 20-40% of the total project value. The remaining finance is usually provided by a variety of debt instruments including commercial bank debt, long term bonds, tax exempt obligations, commercial paper, municipal bonds and industrial development bonds. The limited partners are generally individuals with high incomes who participate in the project to obtain a tax benefit (investment tax credit, energy tax credit and depreciation) that are attached to the project and it's technology e.g. cogeneration. The general or managing partner is either a technical management service, a vendor or an energy service company with expertise in the energy field which sells it's management and operating services.

<u>Utility financing</u> is another general feature of the North American energy efficiency investment scene. The American utilities make use of a wide range of financing mechanisms, such as;

- direct loans;

- loan interest reduction i.e. loans at below market rates;

- equipment rebates, i.e. reduced prices for energy efficient equipment;

- energy-saving subsidies;

- energy-saving guarantee programmes;

- shared saving contracts through energy service subsidiaries;

- leasing programmes through leasing subsidiaries; and

- project financing.

More than 50% of American utilities are now involved in innovative financing of energy efficiency investments. In general the utilities have concentrated on direct loans or conservation incentives to encourage energy efficiency. Their priorities have been the installation of more efficient air-conditioners; heat pumps and fluorescent lighting. Among the more innovative technologies supported are chilled water thermal storage units for space cooling; gas absorption air conditioning; gas fired commercial cogeneration; more efficient electric motors; improved gas fired furnaces; better burners; and electric induction furnaces.

3. The Barriers to Innovative Energy Efficiency Financing in the European Community.

Several factors have been influential in restricting the more widespread utilisation of innovative financing in the European Community. Among the major factors are:

- lack of finance. In third party financing ESCOs in both the U.S. and Europe have tended to draw their finance from a larger parent company. In some cases venture capital, which is more readily available in the United States than in the European Community, has been used to support the creation of an ESCO. Thusfar the traditional suppliers of capital, the financial institutions, have been reluctant to support the operations of ESCOs. The reasons for this are twofold. Firstly these institutions are unfamiliar with the operation of the third party financing mechanism. Secondly, while financial institutions have a considerable experience in the provision of energy supply project finance they have, as yet, little or no experience in the field of energy saving programmes. However, the risks associated with energy savings, e.g. changes in oil price, are not really very different from risks attending energy supply projects. There is no fundamental reason why financial institutions should not become conversant with energy saving project risk assessment after some exposure. One level of risk which may be rather difficult for a financial institution to quantify is the technical capability of the ESCO. There is, therefore, a confidence gap between the ESCOs and the financial institutions which can only be filled by working successfully together.

- lack of knowledge of the techniques. To date the limited application of the techniques explained above in European Community has been caused by the mechanisms not being widely understood, or even known.

- complexity of contract. Third party financing contracts appear complicated to those disposed to make energy saving investments. This apparent complexity has turned many potential clients away from utilization of the mechanism.

- there are some administrative problems which have restricted the application of novel financing techniques. There has been the example in one Member State, where a decision by the Treasury Department that third party financing contracts entered into by local authorities would be considered as expenditures by the authority for that year, and would therefore form part of the authority's budget. This ruling effectively blocked any third party financing investment by the Member State's local authorities.

4. What Actions Have We in the Commission Taken to Promote the Use of Third Party Financing?

To date the Commission has concentrated it's efforts on assisting third party financing to reach it's full potential in Europe and has taken three actions aimed at achieving this.

A study[3] was completed in 1986 which examined the mechanism in detail, and which looked at ways in which a more rapid acceptance of third party financing could be achieved in the European Community.

One of the obstacles to the expansion of this mechanism identified by the study, the complexity of the contracts, was examined in a second study, and a series of model contracts for third party financing in both industry and buildings have been developed. Each of these contracts is accompanied by a commentary which explains in detail, and in layman's terms, the operation of each clause of the contract.

The results of the study on third party financing, and it's potential in Europe have been published in the form of a book which has been circulated as widely as possible. The results of the model contract study were presented at a seminar held in Luxembourg on Oct. 8th. and 9th., 1987[4]. This seminar was addressed by experts in the field of third party financing, and workshops examined the various Clauses of the model contracts. Over 180 delegates attended this seminar, more than 25 of them came from financial institutions, and they took away with them not only an improved understanding of the concept but also a greater appreciation of the interdependence of user, energy service company and financial institution.

On the 29th. of March 1988, the Commission of the European Communities adopted a Recommendation on Third Party Financing. The Recommendation presents a series of actions which the Commission feels the Member States should implement if they wish to accelerate energy efficiency investment through third party financing.

These recommendations include:

a) The removal of legislative or administrative obstacles to the use of third party financing for energy efficiency investments. In particular those restricting the ability of local authorities to use third party financing.

b) The active promotion of the use of third party financing within the public sector.

c) The establishment of national model third party financing contracts along the lines of those prepared by the European Commission.

d) The encouragement of public or private sector enterprises particularly those involved in energy supply, to play an expanded role by providing third party financing services.

e) The implementation of measures to encourage and promote the provision of third party financing services by gas and electricity utilities, particularly for the tertiary and multiple residential sectors, and for small and medium sized companies.

f) To provide grants to multiple dwellings and smaller industrial or commercial enterprises to defray the costs of audits carried out by reecognised energy services and third party financing companies.

g) To initiate measures to accelerate the creation of third party financing enterprises in the energy field by means of financial incentives such as access to deferred interest loans, direct State equity participation or financial guarantees.

h) To establish comprehensive information programmes designed to promote the use of third party financing for energy efficiency investments in all sectors of the economy.

i) To cooperate with the Commission and other Member States in regular reviews of progress and of possible need for additional measures in this field.

5. Why is Third Party Financing suitable for Cogeneration projects?

Cogeneration projects would seem to present the perfect market for the use of third party financing. Of their very nature cogeneration projects tend to be costly. The average cost of a cogeneration project easily exceeds the 100,000 ECU minimum project cost which most ESCOs require.

In general the cogenerator whether industrial, hospital, university etc. will only become a cogenerator for economic reasons and therefore has no interest in the technological aspects of the project. The running of the cogeneration plant can be left in the hands of the ESCO or a specialised maintenance organisation.

The cogenerator is quite used to the situation of buying energy and paying for it over time. A third party financed cogeneration scheme requires no philosophical changes on the part of the cogenerator. He still continues to pay for his energy as he uses it except that he is the owner of a more rational energy production system.

6. Conclusions

What general conclusions can we draw from this rather cursory examination of the of the energy efficiency investment scene?

Firstly, the achievement of the Council's 1995 energy efficiency objective will require investment on a level far greater than that which is currently taking place. The potential market within the Community for third party financing services has been estimated at 86 billion ECU. The size of this potential market should act as a sufficient incentive for the creation of ESCOs but this has so far not been the case. One must not expect that much of this finance will come from public authority sources. It will therefore be necessary for many of the required ESCOs to be created within the private sector.

There is also a second problem. The period 1973-1986 was one of spectacular achievement in the field of rational use of energy in the Community. Our dependence on oil fell from 62% in 1973 to 47% in 1986 while the improvement in energy efficiency was recorded at 20%. While the energy efficiency programmes undertaken by the Commission and the Member States undoubtedly helped produce this result, the major re-structuring of European industry from the older energy intensive industries to newer less energy intensive industries, and the pressure of energy price increases, were significant factors in achieving the improvement in energy efficiency. In 1987, the restructuring of European industry is almost complete and the more attractive energy saving investments have already been made. In addition the short term view on energy prices is that they will stay relatively low. Since two of the planks of our spectacular energy efficiency performance have now been removed we must concentrate on accelerating investment by introducing novel financing mechanisms. How can this objective be accomplished ?

While there is little likelihood that public authorities will provide the investment funds required to achieve our 1995 energy objectives, Member States do have an important role to play in removing all administrative obstacles to the application of third party financing techniques. This does not necessarily mean that Member States should enact legislation to support financing activities, rather that they should remove currently existing administrative impediments to the proliferation of this novel financing mechanism.

It is particularly important that countries, such as Greece, who are at the beginning of their energy efficiency efforts avail of the benefits of third party financing. There is no question of tapping the sizable market for third party financing without accelerating the rate at which ESCOs are being created. There are several mechanisms which can assist this growth.

Government departments and local authorities should be actively encouraged to pursue novel methods of financing energy efficiency investments with the purpose of meeting the European Community's 1995 objective and, thereby, saving tax-payers money without recourse to the use of public funds. The role of Governments in supporting the spread of novel financing mechanisms needs to be stressed. The Member States must accept their responsibility in achieving energy efficiency targets by stimulating their own Departments to have recourse to novel financing of energy efficiency investments.

There is a considerable role for the Commission and the Member States in making novel financing mechanisms better understood by those who should contemplate energy efficiency investments. This campaign should involve seminars and publications, and should be targeted at the very highest level of decision making in both the public and private sectors.

The European utilities should be encouraged to follow the lead of their American counterparts and to think of themselves more as offering energy services rather than simply energy suppliers. European utilities not only have access to clients but they also represent a repository of vast energy expertise which is as yet untapped in the cause of energy efficiency.

A point which should not be overlooked in the energy efficiency process is the matter of the disparity between the rates of return applied to energy supply and energy use projects. It is customary to accept energy supply projects producing an internal rate of return of, say, only 5% while energy saving projects producing rates of return in excess of 25% are all too often considered uneconomic. Capital budgeting normally operates on the principle of accepting the most remunerative projects first. If this principle were applied to the energy sector there would undoubtedly be greater interest in energy efficiency investments.

Finally, the question of mobilising private capital, and reducing the 'confidence gap' between the ESCOs and the financial institutions is central to accelerating discrete energy efficiency investments, and must be addressed on a Community wide level.

The seminar on third party financing held in Luxembourg in October identified this problem and asked the Commission to address it. The communication on third party finance which was presented this year considered several possible Community initiatives, such as a guarantee scheme and a Community wide insurance to help bridge this confidence gap. Such a scheme should be of very limited duration since once a working relationship is developed between the service companies and the providers of capital the requirement for such coverage action would disappear. However, some such guarantee or insurance scheme may prove a vital and necessary element in mobilising the large amounts of private capital necessary to carry out the energy efficiency investments required to achieve the Council's 1995 objectives. The cost of such a scheme should be small in comparison to the benefits it would stimulate. It should help to establish, at a very moderate cost, a climate in which investment in energy efficiency will be as acceptable as other non-energy investments.

The achievement of the European Community's energy efficiency objectives is a unique challenge. There are many obstacles to the achievement of the Community's 1995 energy efficiency objective and to overcome them will require innovative solutions and a rekindling of our entrepreneurial spirit. While most of the finance must inevitably come from the private sector, the public authorities have an essential role to play.

REFERENCES.

1. <u>Third Party Financing Opportunities for Energy Efficiency in the European Community</u>, Association for the Conservation of Energy, Kogan Page, London 1986.

2. COM(86) 264 final, Brussels, 16 May 1986.

3. Ob cit 1.

4.Brown, I., <u>The EEC Model Third Party Financing Contracts</u>, paper presented at the EEC Third Party Financing Seminar, Luxembourg, 8 and 9 Oct. 1987.

COMPARATIVE ANALYSIS OF THE LEGAL CONDITIONS IN THE NON-EEC INDUSTRIALISED COUNTRIES: DIFFICULTIES AND ADVANTAGES

DENIS DRISCOLL

Faculty of Law. University College
Galway
Ireland

SUMMARY

The interest of governments in cogeneration and alternative energy power production has been reflected in a determination to remove whatever obstacles have existed as impediments to the development of alternative power sources - essentially the legal difficulty of selling independently produced power- and to the obstructionist attitude of the utilities themselves. This paper reviews the legal situation in a number of non-EEC industrial countries: The United States, Canada, Norway, Sweden, Finland, Switzerland, Austria, Australia, New Zealand and Japan. It is in the United States that the greatest institutional changes have been made, through the establishment of a legal framework of enforced cooperation between the utilities and the autoproducers.

RESUMEN

El interés de los gobiernos en la cogeneración y la producción con energías alternativas se ha plasmado en la determinación de eliminar aquellos obstáculos que han existido para el desarrollo de las fuentes energéticas alternativas, -fundamentalmente las dificultades legales para vender la energía autoproducida- y la actitud obstruccionista de las compañías eléctricas. La ponencia revisa la situación legal de una serie de países no comunitarios: Estados Unidos, Canadá, Noruega, Finlandia, Suiza, Austria, Australia, Nueva Zelanda y Japón. Los mayores cambios institucionales han ocurrido en Estados Unidos mediante el establecimiento de un marco legal para forzar la cooperación entre las compañías eléctricas y los auto-productores.

COMPARATIVE ANALYSIS OF THE LEGAL CONDITIONS OF COGENERATION IN THE NON-E.E.C. INDUSTRIALISED COUNTRIES: DIFFICULTIES AND ADVANTAGES

Dennis Driscoll
Dean
Faculty of Law
University College Galway
Galway, Ireland

1. INTRODUCTION

Autoproduction of electricity declined as a percentage of total generation primarily because of the economies of scale resulting from the development of large central power stations. In the United States, for instance, autoproduction had accounted for almost two-thirds of generating capacity in 1900 and by 1973 amounted to only 4.2%. This same dramatic decline was witnessed in the industries of other Western countries. However, recent years have seen a renewed interest on the part of many Governments in alternative power production because of an increasing governmental concern with energy conservation, energy efficiency and security of supply. This governmental interest has been reflected in a determination in some countries to remove whatever obstacles have existed as impediments to the development of cogeneration and alternative power sources. Essentially, the obstacles have related to the legal difficulty of selling independently produced power and to the obstructionist attitude of the utilities themselves. In a number of European Community countries there is considerable government interest in encouraging autoproduction. Britain, the Netherlands and Spain are outstanding examples. But, of course, such interest extends far beyond European Community countries. This paper reviews the legal situation in a number of non-EEC industrialised countries: the United States, Canada, Norway, Sweden, Finland, Switzerland, Austria, Australia, New Zealand and Japan.

The legal obstacles of cogeneration revolve around the ability of the independent producer to sell his power at financially rewarding prices: does the autoproducer have a <u>right to sell</u> to the grid (i.e. into public supply) and the grid an <u>obligation to buy</u>? has the autoproducer the right to sell power to the grid on the basis of fixed tariffs, or must sales be negotiated <u>ad hoc</u>? is there provision for independent review of the price structure? has the autoproducer a right to sell power to third parties? if so, are use-of-system charges independently determined? Above and beyond the legal difficulties, the unsympathetic attitude of the utilities has proved a significant institutional barrier to any possible competition from independent producers: the utilities have the bargaining ability to make interconnection difficult, to buy power at excessively low prices, to sell back-up power at high prices, to refuse to transmit power across their lines to third parties, and so on. The legislation in the United States, in particular, has endeavoured to resolve these difficulties by establishing a legal framework of enforced cooperation between the utilities and the autoproducers. The different national approaches are summarised in Table 1a and 1b.

2. THE RIGHT TO SELL TO THE GRID AND THE OBLIGATION TO BUY

In the case of most countries, a prerequisite for the development of cogeneration is that, at the very least, the cogenerator should be able to sell his power to the grid. Put another way, the cogenerator must have a legal right to sell his power to the grid and the grid an obligation to buy it. In most of the countries reviewed, there is no such right. This is true of Canada, Norway, Finland, Switzerland, Australia, New Zealand and Japan. The right exists to a limited extent in the Canadian province of Alberta because the provincial government has endeavoured to encourage small power producers of renewable energies. The 1988 Small Power Research and Development Act will enable certain designated autoproducers of up to 2.5MW to have a right of supply to the grid (at prices which have been set above the utilities avoided cost, in order to establish what potential exists for independent power production in Alberta). The programme will run until 125MW are connected to the grid, or 31 December 1994, whichever comes first.

In Austria, too, autoproducers have a right to sell to the grid and the grid an obligation to buy if the independent generator produces power for his own consumption and has surplus power. Further, three provinces have enacted legislation to give small hydro plants (under 5MW) the right to sell to their local utilities.

The entitlement to the right to sell to the grid is rather more limited in the Canadian province of Alberta and in Austria. The definition of entitlement is more elaborate in the United States. Under Section 210 of the 1978 Public Utilities Regulatory Policies Act (PURPA), electric utilities have a legal obligation to buy electricity from producers which qualify (Qualifying Facilities, or QFs) under Section 201; and the Federal Energy Regulatory Commission (FERC) rules elaborated under Section 201 define a qualifying small power producer as a producer who generates less than 80MW of power at the same site through the use of biomass, geothermal or renewable resources such as wind, solar and hydroelectric resources. In the case of a cogenerator, the energy use has to be sequential, and the thermal output must be no less than 5% of the total energy output.

3. THE ESTABLISHMENT OF THE PURCHASE PRICE

The autoproducer's right to sell to the grid is the *sine qua non* of successful development of alternative power sources. It would be difficult to imagine their development without such a right. That having been said, probably the single key entry decision relates to the purchase price at which the potential entrant can expect to sell his electricity. Here, the significant questions to ask are, whether the cogenerator has a right to sell to the grid on the basis of fixed tariffs, whether the tariffs are established independently of the electricity supply industry, and whether the methodologies adopted for determining the purchase tariffs fulfill the objectives of the legislative scheme to encourage independent power production.

The setting of fixed tariffs will encourage entry by reducing uncertainty. Potential entrants will know the price they can obtain for future power sales and can therefore make an assessment of the likely future profitability of electricity production. The effectiveness of the price guarantee increases with its duration. A purchase price which is subject to change yearly, as in most Canadian provinces, for example, operates as less of an incentive to potential entrants than one which can be guaranteed for a long period, such as in the state of Connecticut in the United States, where the state Public Utility Commission requires utilities to accept twenty-year contracts.

The effectiveness of the price guarantee depends upon the purchase tariff being set at such a level as to encourage entry, and the tariff will itself be affected by the electricity supply industry's ability to influence it. Therefore, the question of <u>who determines</u> the purchase price is a crucial one. The greater the utility's ability to determine the price, the more is the likelihood that the price can be set at such a level as to deter entry. To take a European Community country as an example, in the United Kingdom cogenerators have complained that the industry's unilateral ability to set the price has operated as a considerable constraint upon entry. The purchase tariff is established unilaterally by each relevant Area Electricity Board. Section 10 of the 1983 Energy Act provides that the Electricity Council (a statutory body composed primarily of representatives of the industry and charged with formulating general policy and advising the Secretary of State for Energy) must merely be consulted; and the Electricity Council itself is simply under an obligation to consult with the Secretary of State as to the broad methods and principles of establishing purchase tariffs. Such a legal regime contributes further dominance to the industry, which already has unusual bargaining strength in any case.

By way of contrast, in Austria the purchase tariff is set by the Federal Ministry for Economic Affairs. In the Canadian province of Alberta the purchase tariff is set by the provincial government acting under powers in its new 1988 Act. And in the United States the relevant state Public Utility Commissions set the purchase tariffs. The introduction of binding decisions by an independent third party goes a considerable way to redress the bargaining inequality between the utility and the autoproducer.

A number of autoproducers in Canada have argued that a central shortcoming of their system is that the Canadian system (with the exception of Alberta) makes the electricity utility industry both judge and jury in the fixing of tariffs and that what is required is the establishment of an independent authority to review tariffs along the lines of the state Public Utility Commissions in the United States, where, as has been seen above, the Commissions mandate what tariffs may be charged. Although, as will be seen below, the methodologies for fixing tariffs vary among the states, and the tariffs themselves vary quite widely, at least the fact that the purchase tariffs are established only after independent review ensures a degree of objectivity which is lacking in the Canadian system.

In the United States the state Public Utility Commissions set the purchase price at the relevant utility's 'avoided cost'. But this seeming consensus does not really resolve the matter because of the contraversies as to how best to establish an accurate avoided cost. In fact, state PUC's have adopted a number of versions of avoided cost. This has meant, to take a single example, that the Utah Power and Light Company, which operates in a number of neighbouring states, had to pay in 1985 2.6/kwh, 3.5/kwh and 4.8/kwh for PURPA power, depending on whether the power was purchased from a Qualifying Facility in Wyoming, Utah or Idaho.

4. THE RIGHT TO SELL TO THIRD PARTIES

The potential for alternative power production will be considerably increased if the cogenerator has the right to sell not simply to the local utility but also to third parties and, further, if the utility has an obligation to wheel (i.e. transmit) autoproduced power along its lines. Third parties, whether other utilities or large industrial users, may be in a position to offer more attractive purchase rates than the local utility; and the existence of third party purchasers is likely to enhance the bargaining power of the cogenerator with his own local utility.

In the case of the countries under review, there is no right to sell to third parties in Canada, Norway, Finland, Switzerland, Austria, Australia, New Zealand or Japan. The right does exist in Sweden, and in New Zealand there is a likelihood that the Government will shortly introduce such a right. By far the most complicated situation exists in the United States. Sections 211 and 212 of the Federal Power Act as enacted by PURPA give the Federal Energy Regulatory Commission limited authority to order wheeling on behalf of cogenerators and small power producers, but only if a number of stringent conditions are fulfilled. The conditions are designed to ensure that wheeling enhances economic efficiency, improves the reliability of the service, preserves existing competitive relationships, and is not an undue burden on the wheeling utility.

The Act expressly provides in Section 211(c)(4) that FERC's authority to order wheeling is not to extend to the power to order a utility to wheel to retail customers. FERC cannot, therefore, order wheeling to large industrial users. What remains is an authority to order utilities to wheel wholesale power. In this regard, how FERC interprets the threshold criteria, and in particular the requirement that "existing competitive relationships" must be preserved, will have a significant effect on the wheeling possibilities opened up by Sections 211 and 212.

FERC has only begun to consider the problem of interpreting the threshold criteria. In the one case decided thus far, SOUTHEASTERN POWER ADMINISTRATION v. KENTUCKY UTILITIES COMPANY, FERC rejected a request for wheeling because it would have resulted in a substantial loss of sales to the wheeling utility. The Southeastern Power Administration (SEPA) had sought an order compelling Kentucky Utilities to wheel power to eight municipalities which were wholesale power customers of Kentucky Utilities. The sales by SEPA would have displaced 18% of the power that Kentucky Utilities at that time sold to the municipalities. FERC held that the existing competitive relationship would not therefore have been preserved and that, as a result, the application for a wheeling order had failed to meet the threshold requirement imposed by Section 211(c)(1) of the Act.

The development of wheeling opportunities under the Federal Power Act/PURPA is likely to revolve around the interpretation of the requirement to preserve existing competitive relationships. It is significant that in the SOUTHEASTERN POWER ADMINISTRATION Case FERC interpreted this in a narrow way. The competitive relationships in question could either refer to that which exists between the Qualifying Facility (whose power is to be wheeled) and the wheeling utility as to the particular customer requesting wheeling or, alternatively, to the overall competitive relationship between the Qualifying Facility and the wheeling utility. FERC adopted the narrower interpretation and took the view that what must be examined is the bilateral relationship between the wheeling utility and the customer to be wheeled to. The Commission held that "the proper way to determine whether existing competitive relationships would be reasonably preserved is to compare that the wheeling utility sells to the customers that are to receive the power ... to be transmitted and what the utility would sell if it were ordered to wheel". This narrow reading has made it exceptionally difficult to obtain an order compelling a utility to wheel to its own full requirements customers.

In part, it seems, as a result of FERC's reluctance to order wheeling, a number of states have now adopted legislative or administrative rules requiring utilities to wheel QF power in the case of intrastate trade. But the rules vary considerably. For instance, nine states (Connecticut, Florida, Indiana, Maine, Massachusetts, Minnesota, New Hampshire, Texas and Vermont) require wheeling of QF power to other intrastate utilities. Three states (Connecticut, Florida and Maine) provide for compulsory wheeling to affiliated companies of the autoproducer; and two states (New Hampshire and Texas) require wheeling to end-users in certain limited circumstances. The different state approaches are set out in Table 2.

To summarise, FERC has no authority to order utilities to wheel power to retail customers; but it has limited authority to order utilities to wheel wholesale power. It has approached its authority under Section 211 rather cautiously if not, indeed, timidly. In part as a result, a number of states have adopted rules requiring wheeling under diverse circumstances.

5. CONCLUSIONS

A legal regime establishes the framework within which the development of alternative power production can be either facilitated or frustrated. The sample of non-EEC industrialised countries selected for discussion here reveals the divergent legal approaches to the promotion of alternative power sources. Issues such as the independent setting of purchase tariffs, the adoption of the methodological bases for their calculation, and the rights of autoproducers to sell not simply to the grid but to third parties, have been resolved rather differently. These divergent national legal decisions can be expected to have direct consequences on the successful development of cogeneration and new power sources.

TABLE 1a

	Does the autoproducer have a right to sell to the grid and the grid an obligation to buy ?	Does the autoproducer have the right to sell on the basis of fixed tariffs ?	Are the tariffs set independently ?
UNITED STATES	yes	yes	yes (Rates are set by the relevant state Public Utility Commission).
CANADA	no (The only province in which there is a right to sell to the grid is that of Alberta The right is limited to certain designated autoproducers. Some provinces have adopted policies of "encouraging" sales to the grid, e.g. Ontario and British Columbia).	no (While there is, strictly speaking, no legal right to sell to the grid, such sales as do occur take place in most provinces on the basis of fixed tariffs)	no (The only exception is the province of Alberta, where the tariffs are set by the provincial Government on the basis of S.3 of the 1988 Small Power Research and Development Act.)
NORWAY	No	No	No
SWEDEN	Yes	See below (The profit is equally shared between the seller and the buyer. The principles of the pricing system are established, therefore, but the price itself is not fixed).	No
FINLAND	No	No	No
SWITZ.	No	No	No
AUSTRIA	Yes/No (Autoproducers have a limited right to sell to the grid. They may do so only where the power is surplus to their own use. In addition, three provinces have enacted legislation giving small hydro plants (under 5 MW) the right to sell to the grid).	Yes (The tariff is aligned to the State Power Board's wholesale tariff in a range of 80 % - 100 % of the energy charge of the wholesale tariff).	Yes (The tariffs are set by the Federal Ministry for Economic Affairs).
AUSTRALIA	No	No	No
NEW ZEALAND	Present : No	No	No
	proposed : Uncertain	No	No
JAPAN	No	No	No

TABLE 1b

	Does the autoproducer have a right to sell third parties ?	Are use-of-system charges independently determined ?
UNITED STATES	Yes/no (The situation is complex. The Federal Energy Regulatory Commission (FERC) has limited authority to order wheeling, but only on the basis of the fulfilment of a number of stringent conditions. Some states have passed laws mandating wheeling to 3rd parties).	Yes
CANADA	No (In one province, Ontario, the provincial utility will wheel an autoproducer's to its affiliated companies).	Not applicable
NORWAY	No	Not applicable
SWEDEN	Yes	No
FINLAND	No	Not applicable
SWITZ.	No	Not applicable
AUSTRIA	No (However, an autoproducer can sell power to affiliated companies).	Not applicable
AUSTRALIA	No	Not applicable
NEW ZEALAND	Present : No	Not applicable
	Proposed : Yes (It is anticipated that future legislation will provide for compulsory wheeling).	No (However, Electricorp, the State corporation which runs generation and transmission, has agreed to develop a common tariff for use of its transmission grid, which will apply to all users, including itself).
JAPAN	No (However, an autoproducer can sell to 3rd parties if they are within the same building complex).	Not applicable

TABLE 2

REGULATORY AUTHORITY REGARDING WHEELING

	ANOTHER UTILITY	AFFILIATED COMPANY	LARGE INDUSTRIAL USER
Alabama			
Alaska			
Arizona			
California			
Colorado			
Connecticut	x	x	
Delaware			
Florida	x	1	
Georgia			
Hawaii			
Idaho			
Illinois	2		
Indiana	x		
Iowa			
Kansas			
Kentucky			
Louisiana			
Maine	x	x	
Maryland			
Massachusetts	x		
Michigan			
Minnesota	x		
Mississippi			
Missouri			
Montana			
Nebraska			
Nevada			
New Hampshire	x		3
New Jersey			
New York			
North Caroline			
North Dakota			
Ohio			
Oklahoma			
Oregon			
Pennsylviana			
Rhode Island			
South Carolina			
South Dakota			
Tennessee			
Texas	x		x
Utah			
Vermont	x		
Virginia			
Washington			
West Virginia			
Wisconsin			
Wyoming			
	9	3	2

COGENERATION FINANCING AND LEGISLATION IN E.E.C. AND THIRD COUNTRIES

DISCUSSION

SUMMARY

PARTICIPANTS

The following participants have asked questions or made comments:

PERRIN,P., Atochew (France); GREEN,D., Combined Heat and Power Ass. (U.K.); KAUPPS, Ivo; RIVERA, Petroquimed (Spain); PERIS,R., Catalana de Gas (Spain); KOSTIC,D., Comprimo (The Netherlands) and AGUAS,M., T.I.L. (Portugal).

SPEAKERS

Answers were given by:

HAMRIN,J.G., Independent Energy Producer Ass. (U.S.A.), FEE,D.A., Directorate-General for Energy (C.E.C.) and DRISCOLL,D., (I.E.A.).

TOPICS DISCUSSED

-Pay-back time of cogeneration gas turbine systems in the U.S.A.

- New air pollution requirements vs 15-20 years contracts in the U.S.A.
- CEC's requirements for cogeneration financing.
- Price guarantees in the 15-20 years contracts in the U.S.A.
- The link between thermal and electricity prices.
- Examples of utilities operating as ESCO's in Europe.
- Standarized third financing contracts.
- Economic reasons for the development of cogeneration in California and Texas.
- The Economics of cogeneration based on steam revenues.
- Hourly profile of purchasing and selling utility prices.
- THERMIE programme.

COMMENT

The economics of cogeneration and the relationships between the cogenerator and the public utility were the main subjects of discussion. The general experience is that the public utility is paying the cogenerator at least (and many cases, well above) the avoidable cost of the production of that kWh with new capacity. Other relevant points to make are that Spain is the only country in Europe where a public utility is operating as an ESCO and that anyone willing to get a standarized third financing contract should ask for it for the D.G. XVII of the E.C.

ROUND TABLE ON COGENERATION AND ENVIRONMENT

Chairman : J. SIRCHIS

Directorate-General for Energy

Commission of the European Communities

ROUND TABLE ON COGENERATION AND ENVIRONMENT

CHAIRMAN

SIRCHIS,J.,Directorate-General for Energy, Commission of the European Communities.

SPEAKERS

DIAZ VARGAS, A., Directorate-General for Environment, Ministry of Public Works(Spain); DRISCOLL, D.,(I.E.A.); FEE, D.A., (C.E.C.); GREEN, D., Comb. Heat & Power Ass. (U.K.); GYFTOPOULOS, E., M.I.T. (USA) and HAMRIN, J.G., Indep. Energy Producer Ass. (USA).

Opening words by chairman:

Ladies and gentlemen, it is impossible at present speak about the energy policy or about building an industrial plant without keeping in mind the necessity to protect the environment and the existing or coming environmental standards and rules. As far as the European Commission is concerned I should like to mention that there exists a General-Directorate for Environment which deals with the General European Policy in the field of the Environment and also with all the tasks related to the setting up of norms and standards. In addition to this General Directorate there also exists a General Directorate for Research and Development, which develops programmes for novel technologies and novel techniques for improving the quality of the environment and for the reduction of emissions coming from domestic heating, transport and industry. As far as the General Directorate for Energy is concerned, Mr. Fee, who will be the first speaker at this round table, will explain the links between the Energy Policy and the environmental constraints. But I should like to refer solely to the THERMIE programme which Mr. Fee will speak a little bit more about and which he mentioned during the last session. This THERMIE programme includes technologies for reducing emissions using technologies which consumes less energy than the existing ones. This means, this is, another example of the initiative taken at Community level and of the interest the Community has in environmental problems. There will be six speakers, and each of them will speak about specific subjects. The first speaker, as I mentioned, is Mr. Fee, who will speak about "Third party financing and the Environment". Mr. Fee, please.

Mr. FEE: Thank you Mr. Chairman. The topic which I was speaking on this morning is a general one, is a conceptual one. Third party financing is a concept. It is not a concept which is aimed towards cogeneration, it is not a concept which is aimed towards solely energy savings. It is a financial mechanism. It is something you can use in order to carry out a certain project. It depends on whether there is some quality which is measurable at the beginning, which can be saved during the life of the project, which leads to a reduced costs flow and which can be fed back to someone who is willing to make and investment in that project. Most of the projects that are related to cogeneration are very profitable. That is why cogeneration is a very ideal subject for third party financing. But third party financing can also be for environmental projects. Some of the projects we have seen today using bio-mass, using urban waste, have got examples in Europe bringing together the concepts of cogeneration, third party financing and the environmental concerns. Just to give you one small example in the city where the European Community is located, in Brussels, as with a lot of the major cities in Europe at the moment, there is a problem in disposing of urban wastes. So the city of Brussels has got together with a company, which is a third party financing company which has financed a power station which utilizes the urban wastes from Brussels, and feeds the power to the local utility and the steam to some local industries. The basis of this project is that the company, which is generating the power is a financial company with engineering skills. The company which is taking the power are being obliged to take the power by the city of Brussels who are paying the third party financing company the avoided costs of bringing the urban wastes to a pit: the land fill costs. So, here is a type of project which is bringing together the concept of cogeneration, third party financing and the environment. Also, we have got, as Mr. Sirchis said an overriding concern at the moment in the Community with the environmental problem and the Environmental Directorate General has produced a paper which has signaled out energy as one of the major sources of pollution. Next week we will have a paper produced by the Director Generale for Energy, our own Director Generale, which, we hope, will be presented in the middle of next week to the energy working group. Which will point out energy savings as a major priority. There are several reasons for this. First of all, energy savings acts quickly, in other words, if we start energy savings measures today we will have results tomorrow. If we start to build efficient power stations today we will have results in seven years time. The impacts of energy savings is very, very quick. The problems of energy savings in most companies, in most industries, in most facilities in the public sector is lack of awareness on the part of energy managers and secondly, lack of finance. Lack of will and, as someone said today, political will is everything. We realize in the Community that political will is everything. We have at our disposal the technology. We have in the Community an Energy Demonstration Programme that has been running in Europe for ten years, which has given subsidies to 1.300 projects. I think for a total of 1 billion ECUS. A lot of money!. Which has gone into the developing of first class european energy efficient renewable and clean technology. We have the mechanisms for financing. Such as third party financing. We have in Europe large reservoirs of private capital which need to be tapped. What we do not have, at the moment, is the political will, and we don't have the knowledge to put all of these things together. It is very grateful that in Spain, at the moment, the utilities are involved in third party financing activities and eventually these will lead to less construction of power stations which will have, obviously, local environment benefits. But Spain is the only country in the Community, at the moment, who is doing this. There are not another utilities involved in the energy services business. Until we get together and bring the technology, the finance and, finally, the political will, to do something about energy savings, then we will not have the improvements in the environment which has been claimed for us by our Director Generale for the Environment. Hopefully our Director Generale for energy's paper which is being produced next week will spark the Energy Ministers to consider the energy savings as a viable topic if not because of the fact that we need to save energy, in these times of low oil prices, than at least for environmental issue. Thank you.

CHAIRMAN. Thank you Mr. Fee. I would like to say that after all the six speakers we will have time for the audience to make comments and ask them questions. The second speaker at this round table is professor Gyftopoulos who will speak on "What the future of the environment might be without cogeneration" thank you.

Mr. GYFTOPOULOS. Thank you Mr. Chairman I would like to cast my remarks as four separate issues.

First, I would like to tell you how delighted I was to hear about the progress that has been made in Spain in the area of cogeneration. I lectured for several days to spanish industries about the benefits of cogeneration fifteen years ago, as an invited speaker of the Instituto Tecnológico de Postgraduados, as I recall. And at the time there was hardly any effort in cogeneration and it was a delight to hear today the progress that has been made. The second point that I want to make is, as far as efficiency is concerned, when one calculates correctly, I may add, the efficiency of energy utilization by industrialized nations one finds that this efficiency is between 12 % and 15 % on the average, It is very low and therefore there exists tremendous opportunities for improvement. To be sure it will never be 100 %. But nevertheless there is plenty of room for improvements. Having said that, however, since improvements must always be cost effective I must add, the energy problems of either the advanced or the developing countries can not be resolved by addressing only the cost effective energy utilization aspect of the energy equation. We need and we must also to develop a major new energy source and therefore we have to approach the problem both from the point of view of new energy sources as well as from the point of view of better utilization of the sources that we have or renewable resources that we may develop in the future. Mr. Fee this morning made a very interesting statement. He started his remarks by reminding us of how much money will be required to invest in order to achieve certain savings of energy within the European Community. And as I recall his numbers they were in the tens of billions ECUS or something like that. At face value, these numbers suggest that a large investment is required in order to achieve energy savings and if one does not thinks carefully about the problem one might be tempted to assume that this is an expensive proposition. The only way one can pass such a judgement, however, is to compare these type of investment with that would have been required in order to achieve the same energy services by developing new energy sources. And under the present circumstances one would have found that the investment required for the known new energy sources would have been much larger than the estimate of the European Community, for providing the same services with cost effective energy utilization. And for that reason the numbers that Mr. Fee quoted are not exorbitant. They serve literally our best interest in the realm which they can be applied. Finally, the fourth point that want to make is how all these remarks are related to the environment. I like to oversimplify the problem by saying that the cost effectiveness of any activity in our society, any impact on our environment, literally depends on the amount of resources that we use. How many Tons of whatever thing we have to dig from the ground, process, transport, install, maintain and so on. To the extent that we achieve cost effective energy utilization by using better equipment, and I underline the cost effective aspect, invariably and on the average, that implies that we are using less materials, less Tons of materials. And the fewer tons of materials we use the lesser the burden on the environment and for that reason cogeneration as one part of this effort of cost effective energy utilization is a very good prospect and has been a very good prospect in protecting the environment.

Thank you.

CHAIRMAN. Thank you professor Gyftopoulos. The next speaker is Mrs. Hamrin who will speak about "Cogeneration: the way in the environmental transition". Thank you.

Mrs. HAMRIN. Thank you Mr. Chairman, from the environmental stand point, particularly, because of their quality reasons, it would be nice, just from the environmental view, to stop burning all fossil fuels... now!. But that is really not practical and it is not economic. Most countries have some fossil fuel resources. And it is a resource, and none of us want to give up the availability of that resource right away. However, there are ways we can use resources more efficiently and if you remember the chart that I showed earlier and that is in my paper, for the same fuel, say coal, by using coal in a cogeneration mode, instead of a straight coal burning power plant, you can save approximately a 100 tons per million BTU's. So, just for fuel efficiency we can use it more efficiently in a cogeneration mode. I think that, what we would find is, if we can choose the most efficient and the least polluting of the technologies and the fuels that are available to us, in our particular country and our particular State, that that will be an important step as we transition into a more environmental least sound energy generation era. At the same time, as I mentioned earlier, I think it is absolutely necessary that we remember that there is a cost associated with that. The cheapest thing to do is to burn the cheapest fuel, which unfortunately is usually the dirtiest, with no-pollution control. So, if I build a coal plant and burn the dirtiest coal which, is cheapest, because there is not as big demand, and burn it without any control devices, that is going to be economically very cheap, but, environmentally, very expensive. And to pretend I pay for the electricity out of one pocket, but I am paying for the environment out of another pocket, and that they are different things is not practical. We need to remember we have both of those costs. We can actually calculate what it would cost us to reduce the ton of emissions per billion BTU's, by using cleaner fuel and control devices or by planting trees in the case of carbon dioxide. We have data. We know how many acres of what kind of trees it would take to offset a ton of carbon dioxide and we know how much of the different kind of pollution are put out by different technologies. Therefore we can start to reflect in the value of the electricity the benefits that are coming from the environmental side. We should be willing to pay a little bit more for cleaner technologies, now polluting technologies. It is not easy. None of this is easy. If it were easy we would not be here talking about it. But it can be done and you can place a value. So, I think that cogeneration will be extremely important. Specially in the transitory period when environmental issue have become of great concern, but we still want or need to burn fossil fuels. But even in that period we should reflect and have incentives for the cleaner fuels and the more efficient technologies, because they do cost us more to build and to operate . Thank you.

CHAIRMAN. Thank you for your comments Mrs. Hamrin. The next speaker is Mr. Green, and he will speak about "Cogeneration and the environment. The energy efficiency approach":

Mr. GREEN. Thank you very much. In looking the subject one area that I think we have to bear in mind it is not only the contribution cogeneration plant can make, when is new, to improving the environment, but also the scope there is for reducing emissions from existing plant, and from existing multiplicity of sites, through, not only using cogeneration in industry, but exporting their heat, particularly in the urban environment, through the district or community easing system. There have been a number of that studies there have been done in the past year or so, since the green house issue really began to take off in the media and in the political sphere, which indicated that your are going to make substantial savings on green house gas emissions, particularly CO2 emissions, by going the cogeneration route. Some of those savings would come because of improvements in the technology through gas combined cycle plant. But one area that, I think, does need to be considered and looked at is, for example, old industrial infrastructure where you have got an old urban area. You probably are heating those industrial sites by old more inefficient pooled systems. If you look at an urban area you may have a multiplicity of individually heating systems. Some areas will be solid fuels (coal) others it would be gas; quite often it would burn in over-inefficient appliances and they, individually, will add up to quite a lot of collective pollution and increasing CO2 emissions and therefore greenhouse gas problems. Some of the studies, that I have seen recently, indicated that it is technically feasible through a good cogeneration district heating route, for example, to cut down greenhouse gas emissions by something like 30 %, Thus, technically feasible, in other words, economically feasible. And it will not be difficulties in achieving that. However, in this area we have looked at, particularly, in an old urban area and also in the future in areas where we are going to be going for new buildings... I mean, in Madrid for the past few day I have seen that a lot of new building is going on in the city... the same in other european capitals, a lot of new building, industrial structures being created and infrastructure work going in. That is, an opportunity to see how you can provide heat and power to these new areas in a non polluting way to looking at you can reduce your emissions from day one by going the cogeneration route and by improving end use efficiency as well. It is not good all of us producing environmentally acceptable heat and power from our cogeneration stations and put them into a industrial plant or new building that is poorly isolated, poorly control, have inadequate management system... it is the total package we are talking about. It has been mentioned before that cogeneration is very much a route into environmental transition. When I was suggesting things from my few words now I see that cogeneration and its link to energy efficiency, in total terms, it is one thing that could happen, but immediately. And also the solution in the longer term for the greenhouse gas and another environmental problems we have to be facing... and I feel through the combination effort on end use efficiency and production efficiency we can being to tack on some of the problems which are already well known about and which whom Mr. Fee was saying occupy the mind of Council of Ministers in the not long distant future. Thank you.

CHAIRMAN. Thank you Mr. Green. The next speaker is Mr. Driscoll who shall speak of "Governmental interest in environmental issues and alternative energy sources.

Mr. DRISCOLL. Thank you Mr. Chairman. Well, I have only three rather brief points to make in this regard. The first is the concern in the western world, or to put in other way the member states of the International Energy Agency. It seems to me with regard to the issue of the environment and energy different rules ought to apply to environmental attractive technologies and, to some extent, states are beginning to develop such different rules, Let me just give you three examples. I have mentioned rather ellipticaly, a few minutes ago, in my presentation, that with regard to the State of Texas, for instance, autoproducers can sell to the grid, if the autoproducer is less than 10 MW energy using renewable resources. Which it seems to me that makes a great deal of sense. To take a second example the problems of Alberta in Canada as of last year, which wants to encourage renewable energies. They have encouraged the autoproducers to contribute a 125 MW to the grid over the next couple of years and they have set the price level well above the avoided cost in order to encourage what renewable sources there may be in the province of Alberta to see what in fact will happen. Spain did the same thing some years ago by setting the own price level above avoided cost. I do not really see difficulty with that, it seems to me perfectly legitimate to set the price level above avoided cost in order to develop attractive, renewable resources technologies. I think the utilities themselves have got to be compensated in some fashion from buying at that price level but that is a different matter. So, the first thing is that, I think, rules can be develop to encourage environmentally attractive technologies. The second is the prices of the power being sold. We were talking once o twice earlier today about the price paid to a cogenerator for his power and we were saying that it is in a number of countries the avoided cost to the utility, but it seems to me, just as Jean Hamrin were saying earlier that there is another costs which have to be born too, there is the savings in environmental terms, that due is a cost I think is appropriated to pay the autoproducer for that particular environmental cost because it is certainly one we would suffer and it is unrealistic not to calculate that into the equation. The third point to make is that is really rather early days to make comments about Government Policies. Governments have begun to show considerable concern about these very issues but it is fair to say that it is too early to see what policies countries are positively developing. An example will be Sweden. Because Sweden is phasing out its nuclear power it has to turn to alternative sources, but really at this particular stage is not certain quite how to do it. There are no rules yet being developed affecting the behavior of alternative power sources. Well, the swedish example is not unusual. The same is really true in a number of another countries. It will be true in Norway and Finland, for instance. There is a concern to encourage more attractive environmentally sound sources and an interesting encouraging independent power production. But it is not all clear how this is to be done. In some small measures I think that workshops like this are valuable because at least people from different cultures can begin to learn about the experiences of the other cultures and perhaps bring home at least half of an idea which might be worth talking with one's colleges in the utilities, in the Ministries and among companies who may consider independent power production.

CHAIRMAN. Thank you Mr. Driscoll. It is not a coincidence that the following speaker Mr. Diaz Vargas belongs to the Ministry of Public Works and he will probably illustrate the comments made by Mr. Driscoll with regard to the "Energy and Environmental policy in Spain". Mr. Vargas.

Mr. VARGAS. Thank you Mr. Chairman. Good afternoon, Ladies and Gentlemen. I would like first of all to make a comment, a brief comment: to say that environmental policy in Spain is just beginning. It is so much in its initial stages that we will only be able to discuss it, on a sectorial level, where it has its own strategies. The European Community Initiatives are the basis for everything that has been done here, both at regional and national level. Now we have to make an effort to develop a policy of this type starting practically from zero. The world situation is very important above all when there are countries that, let's say are underdeveloped, with large energy resources and important development potential such as the Soviet Union or China, for example, which are trying to develop very fast. This might lead to energy sources that are perhaps not very appropriate when talking about the world environmental situation, specially with regard to the greenhouse effect and other types of wold wide situations. I think that having made this general comment we will see how we are trying to do things in Spain and I think perhaps I should add that we do not always make the policies that also favor renewable energy but I think cogeneration policy is important because it can increase energy efficiency and also decrease the use of energy resources. This of course will certainly affect the environment positively but we also need to favour the use of renewable energy sources, for example, waste which is often considered a possible source of energy. But the main source of energy from urban waste might come through recycling or reintroduction in the productive system rather than incinerating. So that, in this sense, we have a waste policy which seems to stimulate cogeneration and at the same time might lead to certain activities that favour specific technologies. Also a very important topic that should be taken into account in environmental policies, is the question of costs. Here mention has been made of the avoided cost and also mention has been made of what has been done regarding the environment and cost. European Community Environmental policies have gone along these lines. It is not a question of saying that anyone who pollutes and pays for it will have the right to pollute, but rather how much is going to be needed, with regard to cost, to reduce this pollution. So, these costs are often being generated and they affect third parties. If the people who affect the environment negatively are persuaded to use cogeneration I think that this would certainly make it possible to reduce pollution. We need to consider the additional cost involved in the Spanish environmental policy in which we apply the new Community Directive of large facilities. As we are in a country that is burning low calorie energy sources, and the additional environmental cost which this represents should theoretically affect the tariffs. We should favor the small energy producers that exist today. An important aspect of these policies with regard to environment and cogeneration is going to involve instruments, economic instruments, that will support policies in the way that one wants in the case of environment, and act as a disincentive for the use of a technology that goes against these objectives, environmental objectives. Of course what we are doing at the moment is advancing investment in a business that promises to be profitable and cogeneration will be implemented. And we have to say this, because cogeneration is profitable. So, this is a source of financing, which is of course one of the obstacles that also comes up and this is going to help to introduce this technology. The environmental policy that is being designed by the Spanish administration attempts to meet the sectorial policies and to use the existing instruments to apply them in those cases where environmental objectives exist. We are trying to establish prior reference framework or to provide incentives in case alternative energies are more interesting from the environmental point of view as compared with strategies that have been designed according to sectorial policy. It seems to me that taking into consideration this framework we will be able to talk about these questions in more detail during the debate. Thank you.

CHAIRMAN. Thank you Mr. Vargas, During these six presentations many subjects, many ideas have been raised. I think that now it is time for the audience to react, and to say whether they agree or not with what has been said, or to ask the speakers questions. The microphone is at your disposition.

ROUND TABLE : COGENERATION & ENVIRONMENT

DISCUSSION

SUMMARY

PARTICIPANTS

The following participants have asked questions or made comments:
MORENO,C., Unión Eléctrica Fenosa (Spain); MARANIELLO, Ansaldo (Italy); DIAZ-CANEJA,F., Escuela de Minas (Spain); KATOPODIS,G., Asprofos S.A. (Greece)

SPEAKERS

DIAZ VARGAS,A., Directorate-General for Environment, Ministry of Public Works (Spain); DRISCOLL,D., (I.E.A.); FEE,D.A., (C.E.C.); GREEN, D., Comb. Heat & Power Ass. (U.K.); GYFTOPOULOS, E., M.I.T. (USA) and HAMRIN, J.G., Indep. Energy Producer Ass. (USA).

TOPIC DISCUSSED

- Alternative energy sources in the U.S.A.
- New energy sources after the transition period.
- Demand for electricity and energy savings.
- CO_2 emission policy in Sweden.
- Outlook of the CO_2 emissions in Europe.
- Energy saving and the transition period.
- North world vs south world in the energy-environment equation.
- Energy savings and environmental problems.
- Energy savings in the transportation sector.
- Environmental policy in Spain.
- Californian regulations for automobile emissions.
- Methane emissions in land filling.
- Energy supply and the single market in Europe.

COMMENT

It was clear after the discussion that USA is one of the most advanced environmentally conscious countries. There, the lead is taken by the State of California. All the participants and speakers agreed that one of the best ways to improve the environment and to reduce emissions is through increasing efficiency, and one of the best technologies of energy efficiency is cogeneration. In this respect, it may be worth quoting the words of Mrs. Hamrin, in the sense that: 1) there is much much more renewable energy and cogeneration around than was ever thought,in fact the utilities are complaining there is too much 2) they work very well and 3) the private sector is quite interested in being involved because of its high profitability.

COGENERATION IN

EUROPEAN COMMUNITIES'MEMBER STATES

Chairman : Mr A. Ceña

International Relations

and

Marketing Director

I D A E

THE EXPERIENCE OF ONE ENTERPRISE

JAIME JOSE CAPARROS

Papelera del Jarama, S.A.
Velilla de San Antonio. Madrid
Spain

SUMMARY

The Papelera del Jarama's experience can be summarized as:

- To have a cogeneration power station that will be paid by the generated energy savings.
- To have a cost advantage over our competitors due to less energy costs by 1992.
- To have improved the environment.
- Not to have had to spend the company's money in making the investment.

Our company makes paper, we are not energy profesionals. All the goals have been reached due to the colaboration and help of IDAE. The Temporary Enterprises Union has had many problems starting up because there was no previous experience in Spain in jointventures of this type. We feel that this system, which has been impelled by IDAE, is the best way to achieve cogeneration for financial and technical reasons.

RESUMEN

La experiencia de Papelera del Jarama se puede concretar en:

- Poseer una planta de cogeneración, que se financiará con los ahorros de energía.
- Poseer, para 1992, menores costes que nuestros competidores, debido a inferiores costes de energía.
-Haber mejorado el medio ambiente.
-No haber tenido que emplear dinero de la compañía para realizar la inversión.

Nuestra compañía fabrica papel, no somos profesionales energéticos. Todos los objetivos alcanzados, lo han sido gracias a la colaboración y ayuda del IDAE. La Unión Temporal de Empresas ha tenido muchos problemas para su inicio, dado que no había experiencia previa en España de joiventures de este tipo. Este sistema, impulsado por el IDAE, pensamos que es la mejor forma de llegar a la cogeneración por razones tanto financieras como técnicas.

THE EXPERIENCE OF ONE ENTERPRISE

Mr. Jaime José Caparrós
Manager of Papelera del Jarama, S.A.
Camino del Río s/n.- Velilla de S. Antonio - MADRID

1. IDENTIFICATION OF OUR ENTERPRISE

Papelera del Jarama is locationed in the Madrid's community, 25 km. from the capital, in a little village called Velilla de San Antonio, just in the left side of the Jarama's river.

Papelera del Jarama is a paper factory, that can be designated as a medium one inside the spanish context. Its yearly production is 20.000 tons of paper. These tons are destined to the corrugated board sector and its transformation in packing.

The raw material that we use come from the waste of Madrid's city, and from its industrial belt. The waste paper is splitted with water, without using chemical products. By this way we can recover the cellulous to make with it the new paper.

Our factory has an autonomous section of splitting and treatment of the cellulous, with two helicos pulpers, of new technology with 6 and 12 m.; also there are two continous paper machines with an useful wide of 245 cms. The machines give as ending goods the paper reels with a diameter of 125 cms. and a weight near to 2.000 kgs.

There are 45 persons working, 37 of them are working in regime of continuous work, making 4 and 1/3 turns, in teams of 8 persons, 7 are working with the production and the other person is the group boss.

So by this way of working, the factory has 326 days, with 7.824 annual hours.

2. EXPERIENCE BEFORE THE COGENERATION

2.1. STEAM

Until the end of 1986, we worked with two fuel-oil boilers, with a good yield, but also with the normal problems of this kind of combustible, as can be, the supplying, changes in the fuel-oil characteristics, storage tanks to keep it, treatment to low temperatures, filters and cleaning of the steam boilers.

With the arrival of the gas to Madrid we gave up the old burnings that worked with fuel-oil and we got anothers of natural gas. With the gas we had the advantage of having a continuous supplying, cleannes in the installations, regularity with the providing; that is a higher stability in the exploitation of the boilers, so the drying of the paper is better too, and we can have more production.

2.2. ELECTRICITY

Unión Eléctrica Fenosa has been, through his line of 15.000 volts, our supplier of electrical energy. This energy was passed to 380 volts by the transformation station of our enterprise.

As we were dependents of the Electrical company, when the supplying was cutted, by any motive as could be: atmospheric or technical reasons, our processing was stopped with the consequent troubles and economic damages.

2.3. THE ENERGETIC COST

In order to better understand the great importance that the energetic cost has in the paper processing, we are going to give the electrical and steam costs during the year before the cogeneration.

Electricity supplied by Union
Electrica Fenosa............ 61 millions of pts.

Steam produced in our boilers with natural gas provided by Enagas........................ 45 millions of pts.

TOTAL COST DURIN THE YEAR OF 1988...................... 106 milliond of pts.

And to go deeper inside the cost, we have the following:

The energetic cost is, over the total cost of the factory, and without taking account of the amortization, the 14%.

The energetic cost, over the total cost of the factory, but without taking account of the raw and auxiliary materials and the amortizations is the 31%.

3. <u>OUR EXPERIENCE INSIDE THE COGENERATION</u>

3.1. ENAGAS

Our first contact with the cogeneration was in the Seminar of Cogeneration organized by Enagas in November of 1986. By then we had already signed the contract to change the fuel-oil for natural gas.

During this seminar, there was an aplication system to make a viability analysis about conegeration installations.

We gave our datas and after they were processed, the result was positive. As the first analysis was very standard we wanted Enagas to realize another study deeper, which endeed by saying thad we could make cogeneration.

3.2. IDAE

During the first months of 1987 we started the conversations with the IDAE (Instituto para la diversificación y Ahorro de la Energía) to know its opinion and the possibilities to make together the viability study by a specialized engineering.

Since the first moment we have found in the IDAE the necessary support that has made possible the reality that it is today our cogeneration installation.

With all this, we asked for budget to three engineerings, one of them had the job and the study was subsidized by the IDAE with the 44% of the cost.

The job was ended during the summer of 1987 corroborating again the possibility of making cogeneration in our enterprise. The investment project had different choices, moving from 129 to 145 millions of pesetas without financial costs and with an annual saving that could be 34 or 38 millions of pesetas, and with a pay-back of 3'8 years.

A new study was made by another engineering that had built a cogeneration installation, similar to the one we needed. This investment was a little smaller, with a bigger save and therefore with better pay-back than the other.

4. <u>TEMPORAL JOINT OF ENTERPRISES</u>

4.1. CONSTITUTION

With all the results that we had got and with the idea from IDAE and Papelera del Jarama that we could realize with success the project, we determinated the form to colaborate.

So by this path our TJE or JV was born. It was based on the law 18/1982 - 26-05, about the fiscal regime of enterprises association and joinventures, and the industrial and regional societies development. This law was published in the spanish BOE on 9-07-1982.

There was not preceding of Joinventures inside these lands, so we started the study and redaction of the statutes and the electricity and steam providing contract.

By this way, the objective of our TJE (JV) was established, with the acquisition, the installation and the exploitation of an equipment that produces steam and electricity and whose production is sold to Papelera del Jarama for an established period of time that can be modified and that coincides with the necessary to recover the investment.

The TJE (JV) was founded by notarial writing, enclosing another writing of the contract above mentioned. The TJE was inscribed in the special register which is in the Economic and Financial Ministry. We have to say that this TJE has a special form to pay the State. This is by fiscal transparence.

4.2. THE FINANCING

The project financing is made according to the contribution of its members. IDAE has given us the bigest colaboration and financing, so it has a 99% participation while Papelera del Jarama has only the 1%.

Although to answer to this offering by IDAE, Papelera del Jarama decided to give all the exploitation profits to the TJE , keeping the same costs as if it has to work without cogeneration.

With this accord, we have the following:

- Make biggers the TJE profits.
- Reduce the pay-back
- IDAE can get sooner the money invested, so it can be used for others investments.
- Narrow the financial costos by the smaller use of the money in the time.

When the UTE has payed-back the investment, Papelera del Jarama will buy the cogeneration installation by the established price. This price is equivalent to the financial interests of the money used and depending on the time had.

4.3. PROJECT

To elaborate the basic project, we asked for budget to three engineerings, givint the job to AESA (Asesoría Energética S.A., Barcelona)

We had two options to choice the size of the turbine. One was adapted to our electrical needs and the other one was bigger than our necessities so it was necessary to export an important quantity of electricity to Unión Eléctrica Fenosa with relation to our consumption.

We wanted to have a project accordint to the factory needs, so we decided to buy the small turbine of 1 megawatt, from SOLAR (USA).

Our cogeneration central is doubly connected to the electrical net having the possibility of exporting or importing whenever we want.

The medium consumption is 800 kw/h more or less, being the maximum demand 1.100 kw/h. We have this level only with tops of high consumption.

The installation is completed with a steam boiler, pyrotubular, with a post combustion burning whose capacity is 8 tons/hour. That is enough to us, since we need 6 tons/hour.

In this project we can find something new as it is the production of hot water used for the disintegration and treatment of cellulous.

The basic project had as investment 146 millions of pesetas. The profits that could be generated yearly were 46 millions of pesetas, with a pay-back of 3,2 years more or less.

4.4. RESULTS AFTER THE PUT ON OF THE COGENERATION CENTRAL

After six months working we can give some technic and economic results.

4.4.1. Technic outcomes

We have seen that the paper production is more steady, vanishing the stops because of deficiency of electrical supply, so normal before the cogeneration. This carries an increase of production. Now when the installation becomes disconnected from the electrical net, our central starts to work in isle giving energy. After it will connect again but without failing the factory's electrical supply.

The steam production is also very balanced with this . The drying in the paper machine is better. The hot water production is letting us a better splitting of the raw materials, and we hope soon to reduce the time of operating, having profits, by the electrical energy savings.

4.4.2 Economic outcomes

The last months have been the months with more high temperatures, so the air, that feeds the turbine, was hotter than other times. So the production is smaller, but with the results obtained we can say that the annual profits will be at the level foreseen in the basic project.

5. SUMMARY

From our experience we can affirm that the objectives loocked for us have been reached.

- We have a cogeneration central that will be amortized by the generated savings and that will belong to us at the end.
- On 1992 we shall arrive to a great reduction in the energetical costs, so with this we shall have a good level of competition, we don't forget that these costs have a great impact inside the paper world.
- Improvement in the environment import.
- We have not used our self investments to get this objectives.

Our factory is addressed to make paper, we are not energetic professionals. All the ends have been got by the colaboration and the ways given buy the IDAE to us.

We can not forget that the realization of the TJE has been very difficult because it was the first in going on.Now there are others that are following our steps.

This system that IDAE gives to the industries, is for us, the best way to arrive to the cogeneration, because of its financial and technical reasons.

All these helps are led to get the best technical end.

We are grateful to all the public and private entities, and to all that persons who have made possible this cogeneration central.

THE COGENERATIVE DIESEL BRESCIA NORD AFTERBURNING EXPERIENCE

MARANIELLO GIOVANNI

Aerimpianti-Ansaldo
Milano
Italy

SUMMARY

In 1982-84 Aerimpianti constructed the Brescia-Nord cogenerative Diesel Power Plant consisting of two GMT Diesel engines of 12.75 MWe fuelled with Bunker c.

The original plant exploited the heat of exhaust gas, water, oil and air to generate heat in the form of saturated steam for the town hospital and superheated water for district heating.

In order to increase the heat production and to increase the overall plant efficiency as well as to improve the charactristics and qualities of the exhaust gas, two additionel afterburning fired boiler were retrofitted to the plant (1986-88).

After one year of demostration operation, satisfactory results were obtained with regard to energy savings and environment impact of the emissions.

RESUMEN

Aerimpianti construyó en 1982-84 la central de cogeneración de Brescia-Nord. Esta planta posee dos motores Diesel GMT alimentados con fuel-oil (Bunker C) de 12,75 MWe.

La central original utilizaba el calor de los gases de escape, del agua, del aceite y del aire para generar vapor sobrecalentado para el hospital de la ciudad y agua sobrecalentada para calefacción urbana.

Con objeto de incrementar la producción de calor, la eficiencia general de la planta y mejorar la calidad y características de los gases de escape, se añadieron dos calderas alimentadas con los gases de combustión y gas natural (1986-88).

Después de un año de funcionamiento de demostración, se obtuvieron resultados satisfactorios en relación con los ahorros energéticos y los impactos medio-ambientales de las emisiones.

THE COGENERATIVE DIESEL BRESCIA NORD AFTERBURNING EXPERIENCE

Maraniello Giovanni
Aerimpianti - Ansaldo
Via Bergamo 21
20135 Milano
Italy

1. The Brescia Nord Cogenerative Diesel Plant

Aerimpianti, an Ansaldo company of the IRI/Finmeccanica Group, has carried out in 1982 the Brescia Nord Diesel Power Plant owned and operated by the ASM, Azienda Servizi Municipalizzati, one of the most important Italian Municipal Public Company involved in electricity and district heating services;

The plant consists of two Diesel engines manufactured by Fincantieri-GMT B550 type - 14 V cylinders, 428 r.p.m. burning Bunker C. fuel, each generating 12750 KWe and 12500 KWt cogenerated thermal power.

The heat is recovered from lubrificating oil, cooling water, supercharging air and exhaust gases, in the form of superheated water for the already existing district heating and technological 18 bar steam for the nearby civil hospital;

The plant is characterized by an high electrical efficiency at different Diesel loads (41,3% at 100%, 40,2% at 50% load) and a cogenerated thermal efficiency of 39,73%. As the heat recovered in the waste-heat boilers from the Diesel exhaust gases (inlet-outlet temperatures of 380-150° C) is a large fraction, abt 65%, of the total thermal power, any inhibition, though partial, due either to blocking or to shutdown caused by excessive fouling of exchange surfaces, will heavily affect the total heat recovery which is potentially possible.

Although the theoretical total efficiency is quite high (abt 81%), some significant limitations exist:

- high fouling in the waste boilers due to the particulates.
- low flexibility since the thermal production is strictly coupled to the electric one.

The need of producing an additional thermal energy for the district heating and to improve the environment impact of the Diesel emissions as well as to get rid of any fouling particulate carbon contained in the exhaust gases, led to the decision to retrofit the Diesel plant with two afterburning boilers (fig.1) placed between the Diesels and the existing heat recovery boilers.

The afterburning is carried out by natural gas combustion, using the free oxygen contained in the flue gases.

Since combustion takes place without adding any external air, a very high thermal conversion is reached, considerably greater than the traditional steam generators.

2. The afterburning boilers.

The basic elements to dimension the two afterburning boilers are the plant thermal power increase to be guaranteed to the district heating and the burn-out of the solid particulates.

An additional thermal power of approximately 2 x 20 MWt was required in the form of 18 bar saturated steam.

The fraction of unburned particles to be eliminated was established to be greater than 90% for the gaseous products and not lower than 50% for the particulates. Since the afterburning boilers had to be dimensioned taking into account these specific requirements, great difficulties were encountered as the specialized literature did not provide exhaustive information. A theoretical combustion kinetics and burn-out model was developed in order to determine the temperature and volume (i.e. the residence time) to design the afterburning. Infact these two parameters highly affect the costs: high combustion chamber volumes mean higher installation costs; higher temperatures mean greater fuel consumption and consequently higher operating costs. Fig. 2 shows schematically the designed afterburning boiler. Two regions may be evidenced. The first region, when the burners are located upside and the combustion flames are headed downwards, houses the actual natural gas combustion. Approximately 45% of the Diesel gases is used and the combustion temperature is 1300-1400° C. In the second chamber the burnt gases are mixed with the residual flow (55%) of Diesel exhaust gases. After this isoenthalpic mixing, the flue gas move towards the heat exchange banks consisting of an evaporator, a tube bundle of two racks and an economizer bundle with finned tubes.

The gas total residence time is 1,5 sec. The boiler equipped with an adequate number of retractable soot blowers, is able to produce the same quantity of heat as provided by 100% Diesel load operation by the use of external feed air when the Diesel is out of service.

The main afterburning technical data (at 100% Diesel load) are:

- Exhaust gas flow rate (inlet)	103000 kg/h
- Temperature (inlet/outlet)	380°C
- Oxygen (inlet/outlet)	14-8% w
- Total volume	150 mc
- Feed water temperature	130°C
- Saturated steam flow	31,8 t/h
- Fuel (methane) consumption	2100 Nmc/h
- Theoretical design efficiency	98,2%

By adding the afterburning an increased plant flexibility can be achieved By separating electrical from thermal power generation the two energy sources become thoroughly indipendent regardless of the Diesel output load.

In addition, even neglecting the benefits of the unburnt particles, there is a significant advantage of generating thermal power in the form of steam with an efficiency which is considerably higher than the traditional conventional boilers.

The energetic analysis of the Brescia Nord Diesel cogeneration plant equipped with the two afterburning boilers yields to:

- Total power entering the system:
  - Q_t = Q_D (Diesel) + Q_a (afterburning) = 30346 + 19959 = 50305 KW_t
  - E_e = Electrical power = 12530 KW
  - Q_c = Thermal power (cogenerated) = 12056 KW
  - Q_b = Afterburning useful thermal power = 19760 KW_t
  - Total efficiency $= \dfrac{E_e + Q_c + Q_b}{Q_t} = 88{,}2\%$
  - Thermal efficiency (heat recovery index) = 63,2%
  - Thermal/electrical power ratio = 2,54

It can be noticed that there is an improvement of approximately 7% in the overall plant efficiency when compared to the Diesel cogeneration plant without afterburning.

3. The demonstrative afterburning operation

The performance of the two afterburning chambers was monitored for a period of one year (from 1.1.88 to 31.12.88) (8784 hrs). Thermodynamic and thermochemical data were collected to assess the actual energy savings and performances. The data collection continued also after some (both Diesel units) shutdowns, either scheduled or not. These shutdowns highly affected the energy saving performance of the whole plant, as the afterburners were operated as conventional boilers with external fresh air during these phases.

3.1 Energy saving

The afterburners operated for a total of 11164 hrs, 5928 hrs as effective afterburners, 4204 hrs with external fresh air and 1032 hrs with air/ Diesel waste gas mixture.

The total mean hourly steam production was 18,72 t/h, the mean hours of operation under maximum Diesel load was 5805 hrs. The mean electrical power was quite low (7,02 MWe) as the Diesels generated electricity at an average load of 55,1%.

The mean overall heat generation efficiency was 95,1%. This value, although underscores by 3% at least the design value related to the steam production under afterburning conditions, is approximately 7% higher than a conventional boiler.

The afterburners proved to be very reliable in generating thermal power. The anomalous and excessive operation with fresh air, caused by two long lasting extraordinary Diesel shoutdows, decreased the energy saving with respect to the project value.

Infact the one trial year operation data lead to an energy saving of 1539 TOE when a comparison with a conventional boiler (efficiency 88,5%) is made.

Normalizing the energy saving with the equivalent theoretical operation time under maximum load condition (2 x 4000 hrs), a value of 2121 TOE/yr can be derived, quite close to the project theoretical value (2270 TOE/yr) mentioned in the EEC-AERIMPIANTI contract.

3.2 Afterburning of Diesel emissions.

The Brescia Nord Diesel engines are fueled with heavy oil, Bunker C which is characterized by high viscosity, large quantities of asphaltene and carbon residual (Conradson index).
This causes the engines waste gases to be particularly rich in unburnt carbon particles and hydrocarbons.

Gaseous, liquid and solid unburnt particles are released GMT tests on B-550 Diesel engines have shown that the particulate emissions vary from 70 to 110 mg/Nmc, CO from 150 to 200 ppm, H_xCy from 20 TO 60 ppm as propane. The solid particles have spherical shape and a Gaussian distribution of the diameter size.

Roughly 70% have diameters smaller than 0,3μ.The mean diameters value is 0,2μm and the deviation is 2μm in the measured log-normal distribution.

The results concerning the Diesel emissions and the burn-out results of the afterburning process, as measured at the Brescia-Nord plant, are summarized below:

TABLE 1

AFTERBURNING OF DIESEL EMISSION

		100 Load		75% Load	
		inlet	outlet	inlet	outlet
CO	ppm	190	0 - 10	140	0 - 5
H_xC_y	ppm	40	0 - 8	35	0 - 8
NO_x	ppm	990	830	960	860
Particulates mg/Nmc		91	18	65	22

At the exit of the electrofilters values lower than 8 mg/Nm3 were measured. It is worth to mention that the Regional environment Authority (CRIAL) imposed the following limits (at 100% Diesel load):

- particulates 40 mg/Nm3
- CO 140 ppm
- NO + NO2 1146 ppm
- H_xC_y 18 ppm

3.3 Model analysis of particulates burn-out.

During the afterburning process each solid particle shrinks progressively untill eventually disappears. The basic relation linking unically the reaction rate with the particulate burn-out can be expressed by the equation:

$$\frac{d\,M}{S\,dt} = -\,q\,(T,\ P_{O2})$$

which describes the carbon mass quantity dM burnt per unit of exposed surface S, per unit of time dt, with a reaction rate q, which is a function of the absolute temperature and of the partial pressure of free oxygen contained in the gases.

It is assumed that the solid particles are spherical, have density r, initial diameter Do with a statistical log-normal distribution f(Do) and burn uniformly remaining spherical until burn-out occurs i.e. shrinking sphere under chemical reaction control process. The particles proceed in the afterburner under plug flow motion.

The mass fraction of the total unburnt particles can thus be obtained as:

$$f = \int_0^{\infty} f(Do)\left(1 - \frac{2\,q\,t}{r\,Do}\right)^3 . \, dDo$$

This relation shows that one must increase either the reaction rate q (pratically by increasing the temperature) or the average residence time of the particulates in the afterburning chamber. Limited data and correlations are available on the oxidation of carbon particles.

The Nagle-Strickland-Constable (NSC) and Lee-Thring and Beer are thought to be the most adequate to describe the carbon reaction rate.

Both correlations evidence the strong temperature - Arrhenius type - dependence.

The combustion temperatures in the Brescia Nord afterburners have different profiles due to the existance of a first, high temperature (1300-1400° C), pre-chamber (18 mc) followed by a second mixing chamber (132 mc) at lower temperature (850-900°C). Two configurations have been considered: a simplified single chamber (total volume of 150 mc ,an inlet gas flow coincident with the total Diesel exhaust gas flow rate, a constant partial oxygen pressure and a given constant volume average temperature) and the actual double afterburning chambers. In this last case the combustion of the first gas flow rate, 45% of total, occurs in the pre-chamber at high temperature and a short residence time (0,2 sec); then it is mixed (having an outlet new f(Do) distribution, and mass unburned fraction) with the residual gas Diesel mass flow (45% of total) and undergo for a longer residence time, approximately 1,4 sec, to a further burn-out in the second afterburning chamber.

Fig. 3 show the results of burn-out parametric calculation for the single and double chambers and NSS carbon reaction rate. The results show that the burnt mass fraction increases sharply in the afterburning volume average temperature range of 800-1000°C. The calculated values are also perfectly consistent with the esperimental data. This also demonstrates how to design+dimensioning procedure adopted for the afterburning boiler proved to be a valid simplification of the complex reality of the carbon burn-out process. The calculations have shown that the division of afterburning in two separate chambers is maximum for low temperatures ($<$ 900° C), whereas at higher temperatures, the burn-out results of the single and double chamber configuration tend to coincide.

4. Conclusions

The experience gained by AERIMPIANTI/ANSALDO in carring out the Brescia Nord cogenerative Diesel plant retrofitted and up-granted with two fired afterburners and the results of one year of demonstrative operation have confirmed that energy saving and environment protection may be simultaneously achieved.

Great plant flexibility as well as afterburning heat generation efficiency over 95% were demonstrated.

A considerable, even better than expected, abatament of particulates, H_xC_y and Co was experienced.

Experimental results confirmed the validity of the burn-out theoretical model.

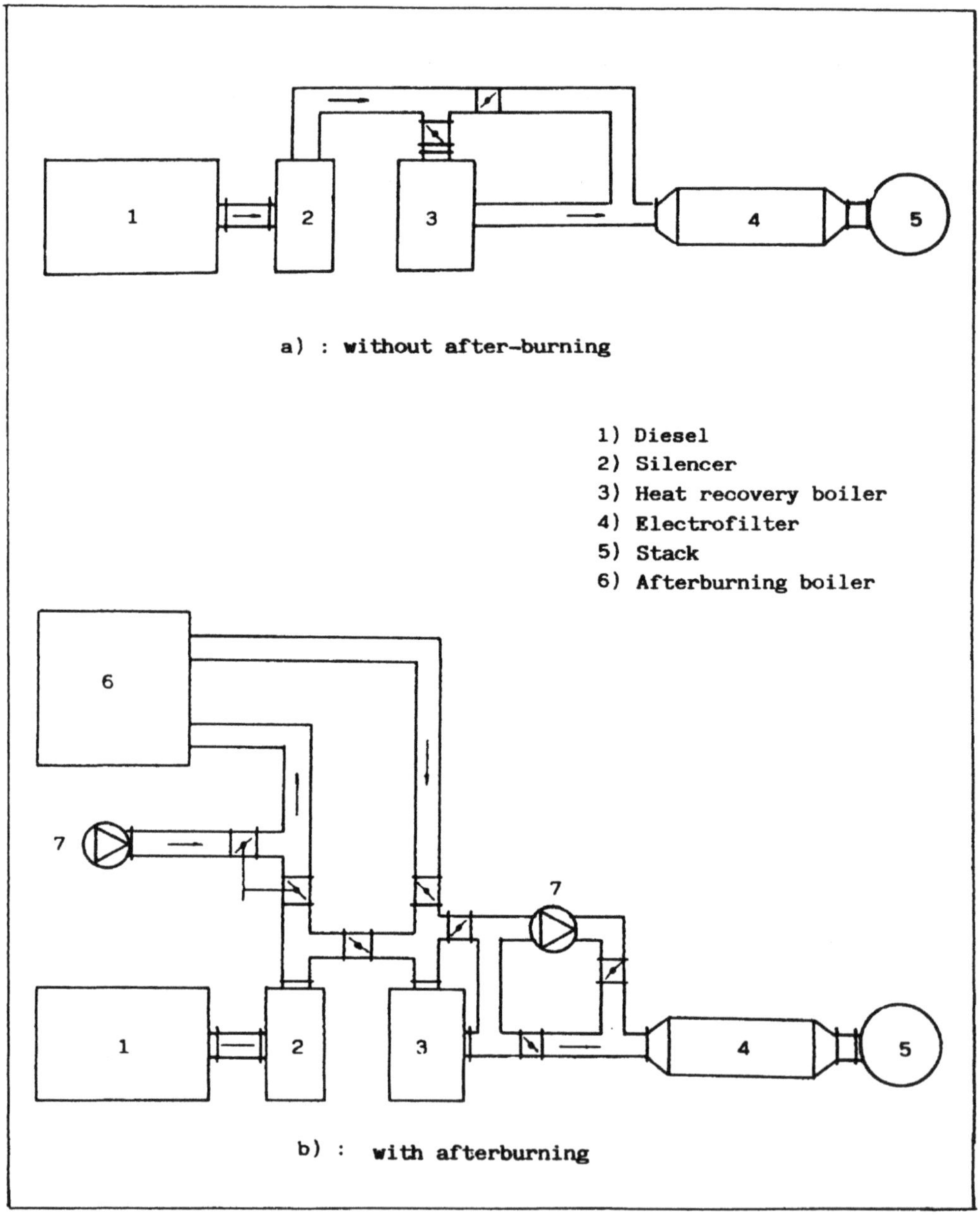
1
2
3
4
5
a) : without after-burning
1) Diesel
2) Silencer
3) Heat recovery boiler
4) Electrofilter
5) Stack
6) Afterburning boiler
6
7
7
1
2
3
4
5
b) : with afterburning

Fig.1

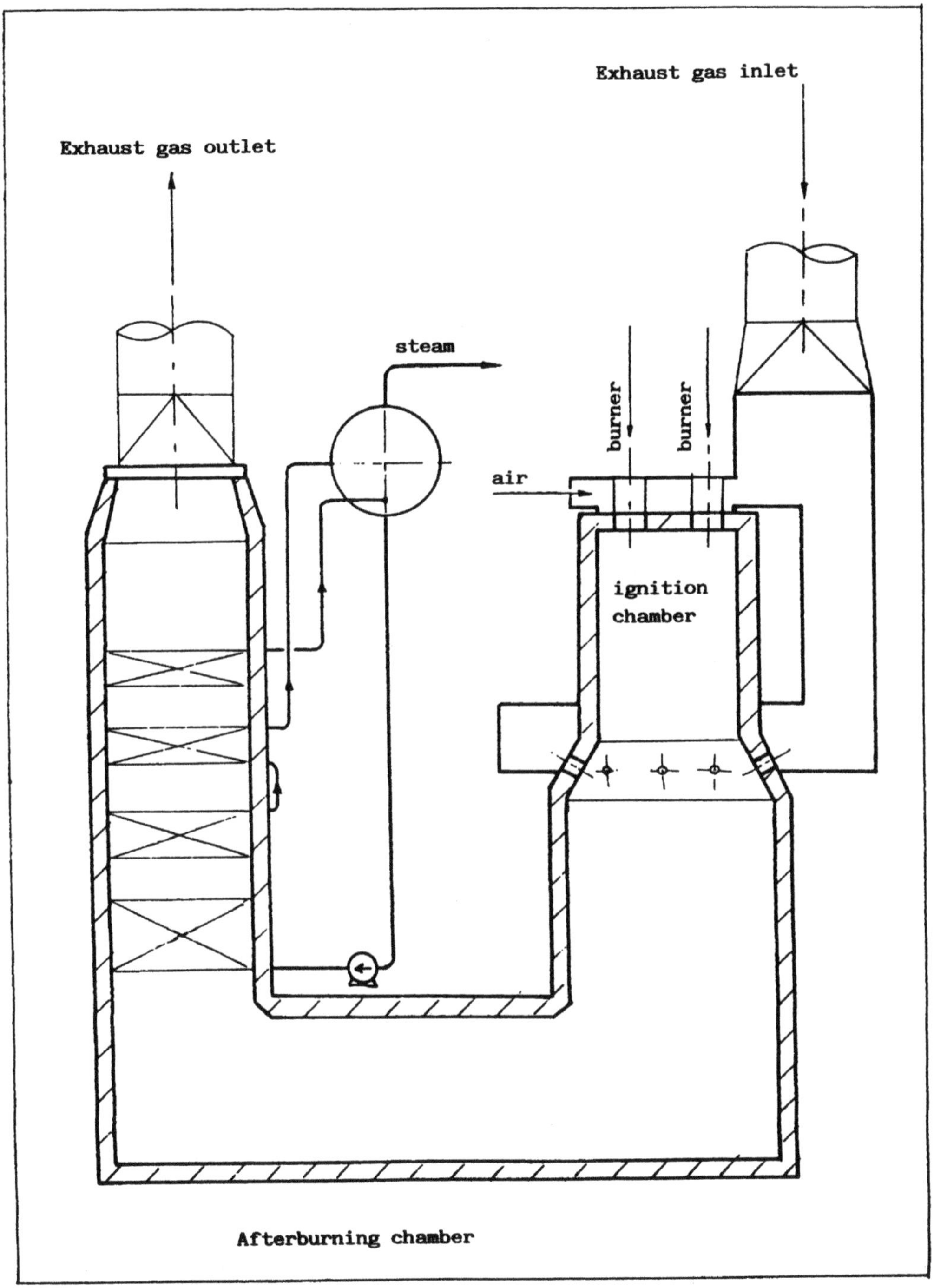
Exhaust gas inlet
Exhaust gas outlet
steam
burner
burner
air
ignition
chamber
Afterburning chamber

Fig. 2

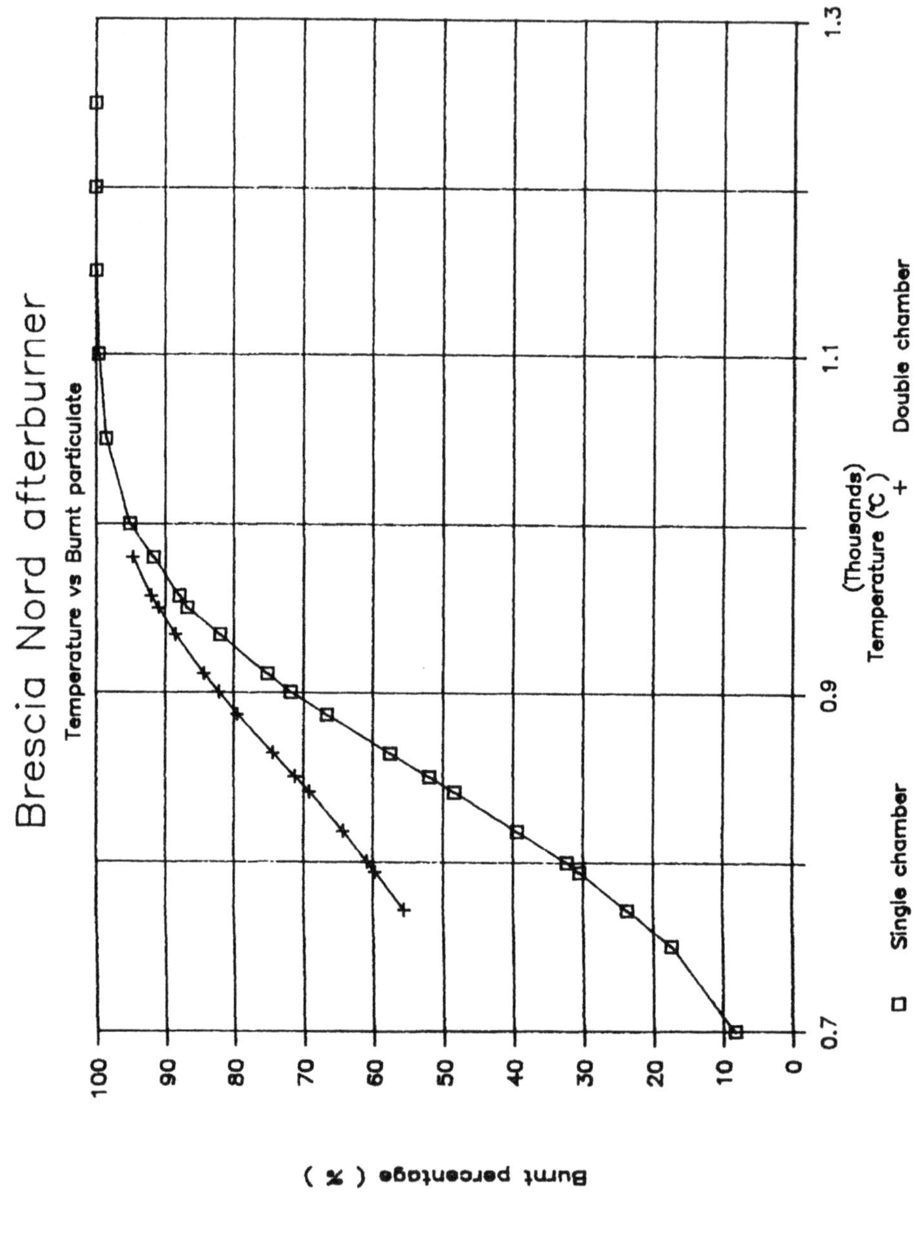
Brescia Nord afterburner
Temperature vs Burnt particulate
100
90
80
70
60
50
40
30
20
10
0
Burnt percentage (%)
0.7
0.9
1.1
1.3
Temperature (°C)
(Thousands)
Single chamber
Double chamber

Fig. 3

MIDLIFE CONVERSION OF A WASTE COMBUSTION PLANT AT DUIVEN, THE NETHERLANDs

F.W. BERKELMANS
P.G. KLOP
F.J. TERMOHLEN

The Netherlands

SUMMARY

The waste combustion plant at Duiven was put into operation in 1975. In 1986 one of three furnaces was equipped with a boiler instalation for the production of hot water, to be used in a district heating system. At the moment, a new energy recovery project is nearing completion. In April 1990 a steam boiler will be put into service, which utilizes the heat from one of the other furnaces. At the same time, an Integrated Energy System (IES) will be made. In this congeneration system the maximum heat from both boilers is utilized directly for heat supply to the district heating system. The heat that can not be used for this purpose is converted into electricity by means of a HP/LP turbine-generator unit. In order to minimize the emission of potentially harmful flue gas substances all 3 furnaces will be provided with a wet flue gas cleaning system.

RESUMEN

La planta de incineración de residuos de Duiven se inauguró en 1975. En 1986, uno de los tres hornos fué equipado con una instalación de calderas para la producción de agua caliente con destino a la calefacción urbana. En la actualidad se está finalizando un nuevo proyecto de recuperación de energía. En Abril de 1990 entrará en funcionamiento una nueva caldera de otro de los hornos. Al mismo tiempo, se realizará un Sistema Integrado de Energía (IES). En este sistema de cogeneración, la mayor cantidad de calor se utiliza para calefacción urbana. El calor que no puede usarse para este fin se convierte en electricidad en una unidad de turbinas (alta y baja presión) y generadores. Para minimizar las emisiones potencialmente dañinas en los gases de escape, los tres hornos estarán equipados con sistemas de limpieza por vía húmeda.

MIDLIFE CONVERSION OF A WASTE COMBUSTION PLANT AT DUIVEN, THE NETHERLANDS

F.W. BERKELMANS, ROYAL SCHELDE, BOILER DIVISION
P.G. KLOP, REGIO ARNHEM, DIENST AFVALVERWERKING
F.J. TERMÖHLEN, PROVINCIALE GELDERSE ENERGIE MAATSCHAPPIJ

P.O. BOX 16
4380 AA VLISSINGEN
THE NETHERLANDS

1 INTRODUCTION

Waste incineration has been one of the means of treating municipal waste in the Netherlands since 1912. Until about 1970 waste combustion plants were purely built for incineration, without any form of energy recovery. However, rising energy prices in the seventies and eighties pushed strongly towards the application of energy recovery techniques in waste incineration installations. In recent years the recovery of energy from waste combustion has become common practice, as in most of the Member States of the European Community.

Nowadays 13 waste combustion plants are operating in the Netherlands with a burning capacity of about 3.5 million tons of municipal refuse a year, including refuse-like industrial waste, representing about 30% of the total combustible waste production in this country. The locations of the plants are shown in figure 1. Details of the installations are shown in table 1.

Nine of these plants, representing about 80% of the totally installed capacity, do already have some form of heat recovery, as for example electricity production, mud-drying, greenhouse or district heating and distilled water production, or a combination of these systems. In the near future the other plants will be provided with energy recovery systems either by newbuilding or by modernization projects.

Besides, all new plants will be furnished with flue gas cleaning installations. In august 1989 the Dutch government readjusted the emission directives considerably, in order to emphasize the necessity of minimizing air pollution. A consequence of the new rules is that within a few years all existing plants need to be extended with flue gas cleaning systems.

2 THE DUIVEN PLANT

The waste incineration plant at Duiven was put into service in 1975. Originally the three furnaces of the installation had been constructed as pure incineration furnaces, that is without heat recovery. During the design stage it appeared, that utilization of the heat being released in the combustion process, was not remunerative.
In order to cool the hot flue gases, the furnaces had been provided with a cooling tower spraying installation. A diagram of the longitudinal section of the original furnace is shown in figure 2.

The plant consists of three furnaces, each with a Düsseldorf-type grate of six rolls. The rolls are 1.5 m in diameter, the width amounts to 3.6 m. The length of the grate in waste transport direction is about 11 m. With these dimensions the original capacity per furnace was 12 tons of waste per hour.

In 1985 a heat recovery project turned out to be remunerative. On one of the furnaces a hot water boiler was installed. The hot water boiler system was integrated in the district heating system of Duiven and Westervoort. As a consequence of the growing waste quantities delivered at Duiven, the capacity of this furnace - furnace no. 3 - was raised to 15 tons/hr. In order to increase the capacity the grate bars were exchanged by a new type. By doing so the primary air flow improved.
In addition, the secondary airflow into the furnace was improved.
At that time the demand for district heating in the region was expected to increase, so adding hot water boilers to furnaces no. 1 and 2 successively would prove to be remunerative.
However, the expected expansion of the two major towns did not take place. Moreover, since the district heating system did not need the full capacity of two hot water boilers for a great part of the year, especially in summer, the fitting of a second hot water boiler would not be economical.

In 1987 a new study was conducted to investigate the recovery of more energy from the furnaces. This study led to a conversion project, which is now nearing completion. The contracts for this project were signed early 1988, commissioning has been planned for March 1990.

3 THE PROJECT

The conversion project concerns the utilisation of the heat from furnace no. 1 by fitting a new steam boiler on it. The capacity of this furnace was raised to 15 tons per hour.
The steam from this boiler will be converted into electricity in a turbine-generator unit. At the same time the hot water boiler and the district heating system will be integrated in the water and steam process. The project will result in a so-called "Integrated Energy System" (IES) enabling a commercial utilisation of all the heat produced by 2 furnaces during the entire year. The IES will supply electrical and thermal energy to the national electricity grid and the Duiven-Westervoort district heating system respectively.

The IES system was designed in close cooperation between the three partners in the project, i.e. Regio Arnhem, PGEM and Royal Schelde.

Regio Arnhem is a co-operative body comprising 18 municipalities in the region around Arnhem in the mid-east of the Netherlands. Regio Arnhem is the owner of the waste combustion plant.

PGEM is a company which generates and supplies electric energy in the provinces of Gelderland and Flevoland. PGEM also manages district heating systems in several areas of these provinces, for instance in the towns of Duiven and Westervoort.

Royal Schelde is a company which employs over 3000 people, which are active in providing a wide range of products and services: steam boilers for industrial plants, power plants and waste combustion plants, naval shipbuilding and repair, process equipment, curtain walling, gear boxes, foundry products and erection services.
The IES is an activity of Royal Schelde's Boiler Division.
The seat of the company is in Vlissingen in the south-west of the Netherlands.
Royal Schelde is the most experienced supplier of power station boilers in the Netherlands (up to 600 MWe).
Royal Schelde for instance supplied the boilers for the large waste incineration plant of AVR in the Rijnmond area near Rotterdam (920.000 tons of waste per year).
Recently Royal Schelde started a long term cooperation with Von Roll, Zürich, Switzerland for new projects in the waste incineration field.

Simultaneously with the conversion project another modernization project is running in Duiven. All 3 furnaces are being provided with a wet flue gas cleaning system in order to minimize the emission of potentially harmful flue gas substances.
The main contractor in this project is Von Roll, Royal Schelde's partner in future Dutch waste incineration projects.

4. THE INTEGRATED ENERGY SYSTEM (IES)

4.1 Process description

In this paragraph the IES will be described, emphasizing the thermodynamical process by passing through a few operation modes.
The main components of the IES will be briefly described in the following paragraphs.
In figure 3 the IES process is shown, comprising the components:

1. Hot water boiler with cooling tower
2. District heating system heat exchanger SV2
3. Steam boiler
4. Turbine-generator-installation
5. Condenser with cooling tower
6. Feedwater preheater
7. Deaerator
8. Steam generator
9. Airheaters
10. District heating system heat exchanger SV1

Items 1 and 2 are part of the existing installation. Items 3 to 10 are new components, constituting Royal Schelde's scope of supply in this conversion project.

The hot water boiler on furnace no. 3 supplies water at a temperature of 180°C to the district heating system via heat exchanger SV2, as in the existing installation. On this point the IES and the original process do not differ fundamentally. However, in the original installation all surplus of heat was absorbed by the cooling tower.
After the conversion the heat surplus of the hot water boiler will all be used for steam generation.
For this reason a steam generator is included in the IES in order to supply 4 bar saturated steam to the low-pressure (LP-)turbine. The feedwater at a temperature of 140°C for the steam generation is fed from the deaerator by two 100% feedwater pumps. The generated steam will expand to a pressure of 0.09 bar in the LP-turbine. After condensing, the condensate is heated to 120°C in a feedwater preheater. Finally the condensate is deaerated by means of steam from the steam generator, which completes this process cycle.
In order to meet a greater district heat demand, the new heat exchanger SV1 is included in a parallel loop. The heat exchanger is fed with 1 bar extraction steam from the LP-turbine; the condensate is passed to the feedwater preheater.
With the hot water boiler in single-operation, serving the steam generator - LP - turbine-loop, a maximum electric power of about 3 MWe will be reached at the generator terminals.

The new Royal Schelde steam boiler on furnace no. 1 supplies steam at a pressure of 40 bar and a temperature of 400°C to the high-pressure (HP-)turbine. The exhaust steam is led to the LP-turbine inlet for further expansion. With this boiler in single-operation mode an electric power of 10 MWe will be generated.

In the future, during normal operation both boilers will be in full operation. Then the electric power at normal steam flow will amount to

13 MWe. At maximum steam flow an electric power of about 15 MWe will be reached. These values correspond with a zero demand of heat of the district heating system.

If district heating is required, this dual operation mode will supply a maximum power of 11 or 7 MWe at district heat demands of 25 or 35 MWth respectively.

Besides, the flexibility of the IES guarantees, that the whole range from zero to maximum district heat can be supplied. Depending on the heat demand at a certain moment, the residual heat will be converted into electric power.

To complete the process, primary air heaters will be installed on both furnaces. These heaters are included in the hot water cycle.

4.2 Hot water boiler

The hot water boiler on furnace no. 3 is a two-pass boiler with convection sections in the second pass. The so-called "membrane" boiler walls are composed of fin-welded tube panels, which form gastight passes. The water flow is controlled by the furnace temperature.

The primary air for this boiler can be heated to a maximum temperature of 140°C.

4.3 Steam boiler (see figure 4)

One of the steamboiler design criteria was the necessity of fitting the boiler within the narrow space between the existing furnace and the electrostatic precipitator. From experiences with other waste incineration installations, we knew that the steam conditions had to be 40 bar and 400°C. In order to satisfy these requirements the steam boiler had to be a two-pass vertical boiler rising up to a height of about 40 m above the furnace grate.

The steam boiler walls are composed of membrane evaporator panels. The dimensions of the first pass are such that the hot flue gases from the furnace will cool down to an acceptable level of less than 690°C before entering the convection (second) pass. The design is without any radiation bundles in the first pass for reasons of corrosion avoidance.

Further the high empty first pass contributes to the burn out of the flue gases, and because of the low gas velocities a low dust content can be guaranteed.

Moreover, light unburnable material (aluminium cans etc.) will not be entrained by the flue gases as a result of this high first pass and the low velocities.

Before entering the superheater the gases pass an evaporator screen which is formed by the connection tubes of the middle wall to the steam drum.

The gases enter the high temperature superheater, which has been designed as a parallel flow superheater. The gas temperature at the entrance is 670°C, thus avoiding high temperature corrosion. Afterwards the gases flow through the low temperature superheater (counterflow). The next bundles are formed by another evaporator and a counterflow economiser. Finally the gases enter the electrostatic precipitator at 200°C.

Both superheaters, the evaporator and the economiser are supported by steam cooled hanger tubes. In order to avoid fouling problems throughout the second pass, a wide tube spacing has been chosen.

The tubes in the second pass are cleaned by sootblowers, injecting superheated steam.

The circulation system of the boiler is based on natural circulation.

The boiler, weighing 250 tons in total, is top supported by means of a newly supplied steel structure.

Expansion joints have been installed at the furnace inlet and rear pass outlet.
The lower part of the first pass is provided with studs, which are lined with refractory. The furnace side walls mainly consist of ceramic insulation in order to avoid fouling. A small part of the furnace walls is equipped with perforated ceramic tiles.

Flue gas flow patterns in the boiler have been studied by means of extensive model research by TNO, an institute for applied scientific research in the Netherlands.
Models have been made of the secondary air flow into the furnace, the deflection of the gases from first to second pass, and the entrance of the electrostatic precipitator. The results of the model research program have been applied to the design.

4.4 Turbine-generator installation and condenser

The lay-out of the turbine-generator and condenser-unit is shown in figure 5. On the foundation are mounted, from left to right: HP turbine - Gear - Generator - Gear - LP Turbine - Condenser
The turbine-generator installation consists of a high pressure (HP-) turbine, a low pressure (LP-) turbine and a generator located in- between. The generator can be driven by one of the turbines or by both operating simultaneously.
The two-different-turbine-mode was chosen in order to obtain a high level of flexibility, since, as we already described in paragraph 4.1, the LP-turbine can be driven by the steam flow from the steam generator, which exchanges heat with the hot water boiler process cycle. This means that electricity can still be produced when only the hot water boiler is in operation and the steam boiler is out of operation for maintenance.
Both turbines are of the axial flow impulse type with horizontally splitted casings.
The two reduction gears are of the parallel gear type. Gravity tanks are installed on top of the gears to ensure sufficient oil supply to the turbine and gear bearings in case of pump failure.
The generator is a four-pole synchronous machine with a built-in brushless exciter and a closed cooling system. The two air/water tube heat exchangers are located on top of the generator.
Steam from the turbine exhaust is condensed in the condenser.
The condensate is led to the lower part of the condenser. The condensate level is controlled and always kept lower than the tube bundle. In order to prevent sub-cooling and waste of energy, the condensate is held at saturation temperature.
Sub-cooling can also cause precipitation of oxygen from the condensate.
The condensate outlet is situated at the bottom of the condenser shell.
The condenser shell is furnished with a connection for air evacuation, vacuum breaking equipment, rupture discs and condensate recirculation. In order to make it possible for the drainage to be collected from equipment such as heat exchangers, steam traps and deaerators, the condenser is provided with a flashbox.
The condenser is cooled by a cooling tower, which uses water that is withdrawn from the ground. The intention is to utilise surface water in the future.

4.5 Process control system

The control system of the IES will be computerised.
This computer system has the following objectives:

- Controlling the fixed and variable process values such as the steam pressure in various systems.
- Monitoring of the process actions such as automatic starts and shutdown of pumps.
- Safeguarding the limit values against overload or underload situations.
- Visualising analogue or binary data by means of graphical displays.

The control system consists of:

- One process computer with extension facilities.
- Two process monitors including the operator communication keyboard.
- One structuring keyboard.
- Two printers.

Special care has been taken to ensure that the system has a high degree of redundancy to ensure a safe and efficient operation at all times.

5. ENERGY SAVINGS

Assuming that the district heat consumption will be as expected and that there will be a regular supply of waste, the IES will be able to provide over 80 million kWh of electricity per year.
In addition, the IES is expected to supply over 60,000 GJ more heat to the district heating system than the hot water boiler in the existing plant. This means a saving of almost 2 million cubic metres of natural gas per year.

6. CONCLUSION

The conversion of the Regio Arnhem plant into the Integral Energy System leads to an optimal utilisation of the energy which is present in the flue gas heat of two waste combustion furnaces.
The project fits in the Dutch environmental policy aiming among other things at large scale waste incineration with energy recovery and flue gas cleaning.
Because of its energy saving and its innovative character the project at Duiven has been selected as an EC-demonstration project.
The modernization project of the Duiven plant will serve as an example for similar waste incineration installations in Europe.

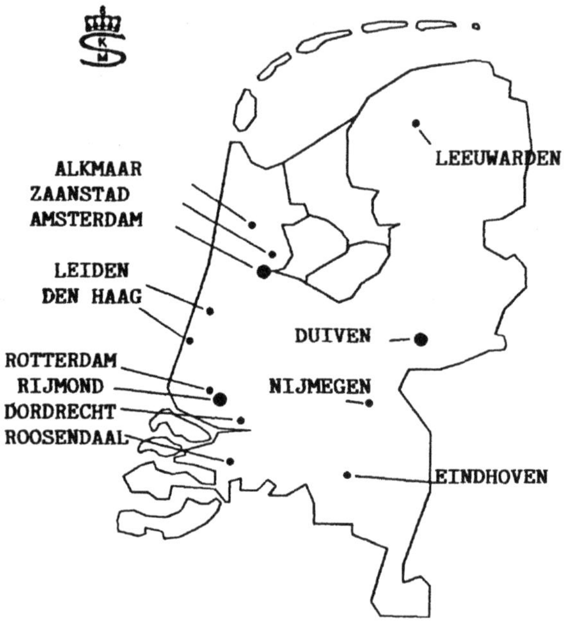

Fig. 1 : Location of the 13 Waste Inceneration Plants

Table 1 : Some details of the waste combustion plants in the Netherlands

Location	Design Capacity [t/h]	Type of Energy Recovery	Year of First Operation	Burnt Waste in 1986 [tons]
Alkmaar	3 x 6	–	1971/1978	112,000
Amsterdam	2 x 16	electricity	1968	395,000
Dordrecht	3 x 7	mud-drying	1972	119,000
Duiven	3 x 12	district heating	1975	218,000
Eindhoven	1 x 6	electricity	1987	–
Den Haag	4 x 12.5	electricity + district heating	1967/1974	285,000
Leeuwarden	2 x 6	–	1973	63,000
Leiden	3 x 4	–	1966/1976	91,000
Nijmegen	1 x 9	electricity	1987	–
Roosendaal	2 x 4	greenh. heating	1976	17,000
Rotterdam	4 x 12.5	electricity	1964	290,000
Rijnmond	6 x 20	electricity + distilled water	1972	920,000
Zaanstad	2 x 9	–	1976	112,000

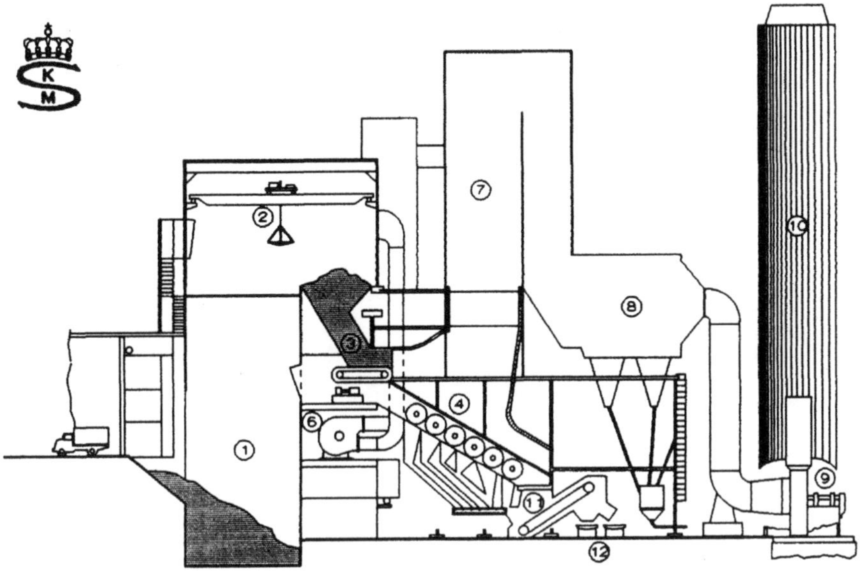

(1) WASTE PIT
(2) CRANE
(3) FEEDING CHUTE AND FEEDER
(4) ROLLER GRATE
(5) FURNACE
(6) COMBUSTION AIR FAN
(7) FLUE GAS COOLING TOWER
(8) ELECTROSTATIC PRECIPITATOR
(9) SUCTION FAN
(10) STACK
(11) WET ASH EXTRACTOR CHAIN
(12) ASH BELT CONVEYOR

Fig. 2 : Sectional Side Elevation of the Original Furnace

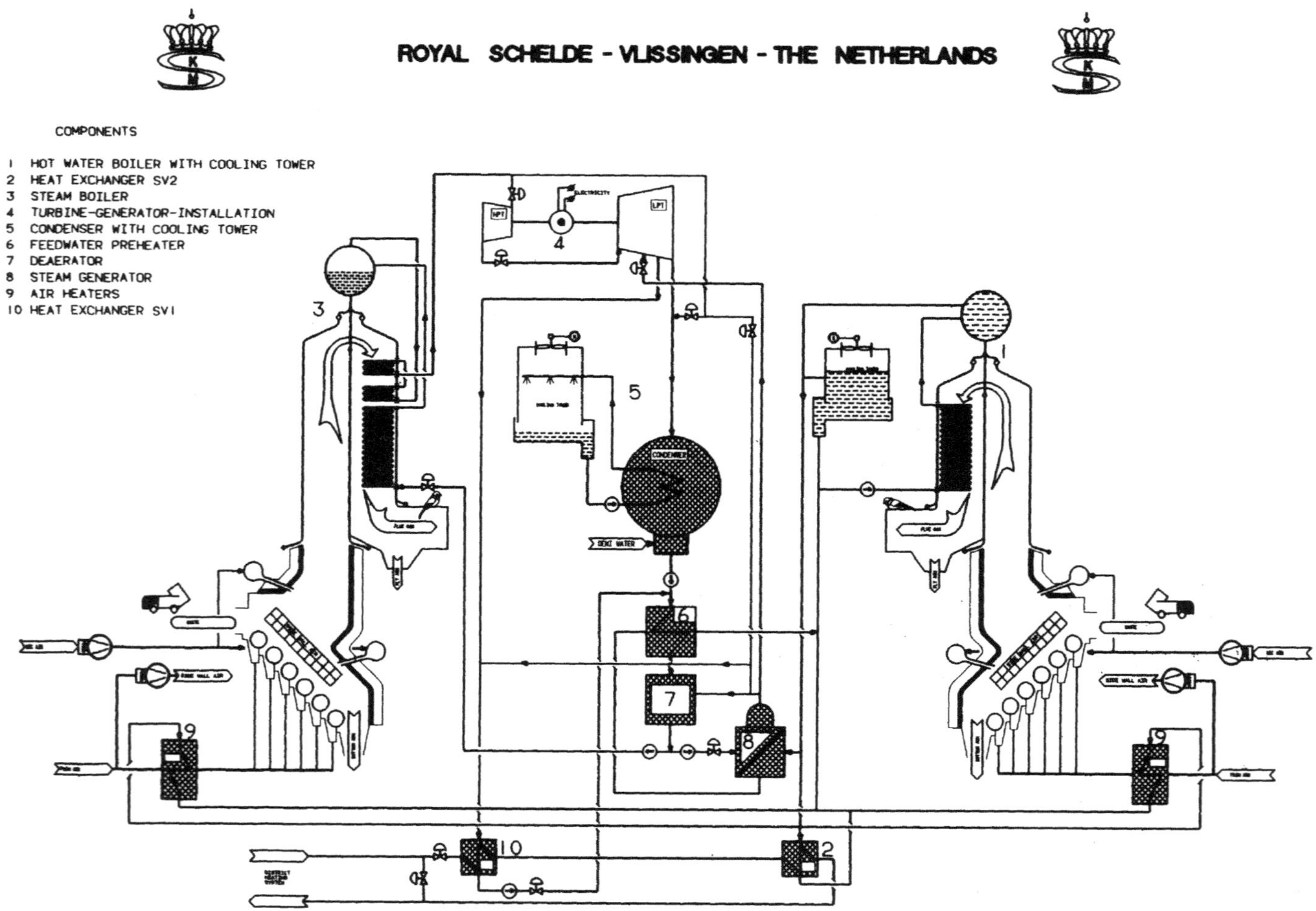

Fig. 3 : Integrated Energy system - Process Scheme

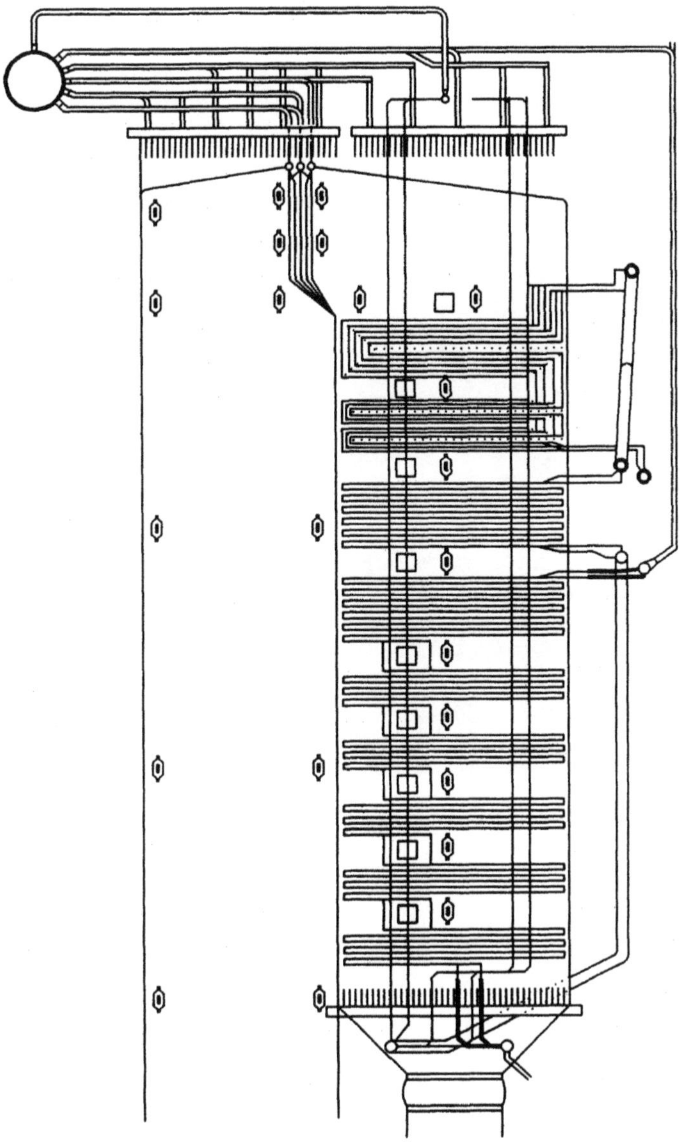

Fig. 4 : Steam Boiler (Schematic)

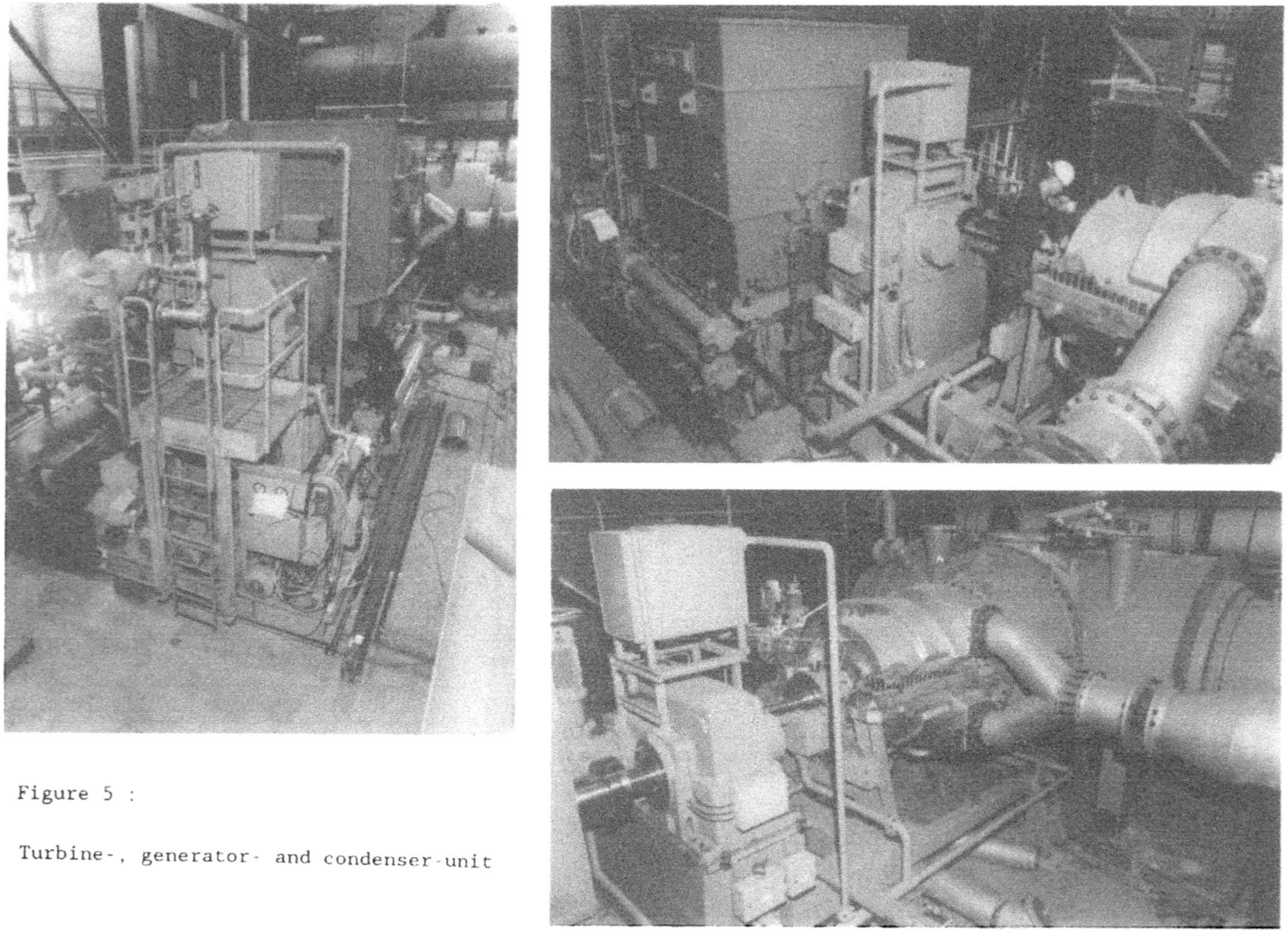

Figure 5 :

Turbine-, generator- and condenser-unit

TECHNICAL AND ECONOMIC ASPECTS OF CHP AT PFIZER

P.P. McGLADE

Pfizer Chemical Corporation
Cork.
Ireland.

SUMMARY

Pfizer Chemical Corporation commissioned a Combined Heat and Power installation at their plant in Ringaskiddy, Co. Cork, Ireland, in August'87. The CHP facility used an existing watertube boiler modified to take the waste heat gases from a 5.4 Mw. gas turbine generator set. A feasibility study carried out in 1985 predicted an efficiency of 91.5 % base on an estimated electricity demand of 37,000 Mwh and a steam load of 302,000 tonnes per annum. Final performance test results indicated an overall efficiency of 87.84 % at 100 % maximun continous rating (MCR) on the boiler and 100 % load on the gas turbine generator set. The boiler was modified to accommodate the gas turbine exhaust gas volume. Savings were somewhat lower than originally estimated due to a fall in both gas and electricity prices but they remain very attractive.

RESUMEN

En Agosto de 1987 la Pfizer Chemical Corporation encargó una planata combinada de calor y electricidad (CHP) para su factoría de Ringaskiddy en Cork, Irlanda. La instalación de CHP utilizó una caldera tubular, ya existente, modificada para utilizar el calor residual de un conjunto de turbinas a gas-generador de 5,4 Mw. En 1985, se llevó a cabo un estudio de viabilidad, estimando una eficiencia de 91,5 % basada en una demanda eléctrica prevista de 37.000 Mwh. y una carga de vapor de 302.000 Tm-año. Las pruebas finales de trabajo dieron una eficiencia global del 87,84 % al 100 % de potencia máxima contínua en la caldera y al 100 % de carga en el conjunto turbina-generador. La caldera se modificó para recibir los gases de escape de la turbina de gas. Los ahorros fueron algo menores de lo que se estimó en un principio, debido a la disminución de precios de la electricidad y el gas, pero no obstante, se mantienen muy atractivos

TECHNICAL AND ECONOMIC ASPECTS OF CHP AT PFIZER

P.P.McGLADE, IEng.,AMIMarE.,CDipAF., Dip.Prod.Mgmt(IMI).

PFIZER CHEMICAL CORPORATION
Ringaskiddy,
Co.Cork.
IRELAND.

1.INTRODUCTION

It is the policy of Pfizer Chemical Corporation to continuously improve energy efficiency at all their plants. As a result of the quest for greater efficiency a study in 1983 looked at the feasibility of a CHP installation but the economics at the time were not favourable. In September 1985 the company commissioned Ewbank Preece Engineering Consultants (Dublin) Limited to prepare a feasibility study on the viability of a CHP development.

The primary objectives of the study were:-

(a) to examine the options for a CHP development utilising a gas turbine generator in conjunction with the existing Blr.3 modified to operate as a waste heat recovery boiler with supplementary gas firing

(b) to identify the optimum arrangement for the CHP Development

(c) to prepare a cost estimate for the optimum arrangement

Previous studies had concluded that a gas turbine was the most suitable prime mover and that the modification of Blr.3 was possible and it would allow advantage to be taken of an EEC demonstration project grant.

The study examined six different gas turbine generator sets and compared the operating costs of these with the cost of single purpose steam generation and total electrical power import. The study concluded that the best option was a Centrax CX571 gas turbine generator set using an Allison 571KA gas turbine rated at 5.4 MW (ISO). An important consideration was that this turbine's exhaust volume was within the capacity of the boiler furnace and fans and therefore maximun heat recovery was possible, other turbine in the desired power range had exhaust volumes which exceeded the boiler capacity and so full heat recovery would not have been possible thus reducing savings.

The overall efficiency of the installation was predicted to be 91.5%(LCV). The boiler efficiency was predicted to be 93.3% (LCV) at 5.4 MW and 93.8% (LCV) on CAM (cold air mode).

2.PROJECT MANAGEMENT AND EXECUTION

The CHP project was approved in March 1986 and a target set to have the system commissioned and on line by the end of the plant annual shutdown at the end of July 1987. The contract was placed for the gas turbine generator set recommended by the feasibility study. It was a contract for design, supply, testing, and commissioning of the gas turbine generator set and its auxiliaries and was awarded to Centrax Gas Turbine Division Ltd, Devon England.

The consultants employed to carry out the feasibility study were retained to manage the project in conjunction with our own in house project team. Twelve separate contracts were awarded for the entire project:

- gas turbine generator set and auxiliaries
- 10kV circuit breaker
- current and voltage transformers for 10kV system protection
- 10kV cable and termination kits
- 10kV /10kV isolating power transformer
- 10kV interface and protection panels
- 415V switch fuses to extend LV switchboard
- incoming and emergency diesel generator circuit breakers for LV switchboard
- electrical installation services including MV, LV and control cabling, earthing and small power and lighting associated with the gas turbine and gas compressor buildings
- civil works for buildings, cable ducts and roadways
- conversion of Boiler 3 to waste heat recovery duty, supplementary firing equipment, turbine exhaust gas ductwork including bypass stack and dampers
- mechanical installation of service pipework for cooling water, compressed air, natural gas and diesel fuel

Work began during the plant annual shutdown in July 1986 on the electrical terminal points and on service pipework terminal points as well as some civil work on relocating underground services away from the planned development.

Civil work began in September 1986 on the gas turbine house and on the gas compressor house and was completed by Febuary 1987 well in advance of the arrival to site of the main equipment.

The gas turbine generator set underwent functional testing and partial load testing up to 3.5 MW prior to delivery in March 1987. Installation of the set and auxiliaries was complete by the end of May '87 and high pressure natural gas was available at the begining of June '87.

Boiler 3 was taken out of service and modifications began in March '87, pressure parts modifications were complete by the end of May '87 and the burner system was tested during June '87.

The CHP system was commissioned and on line in August '87 running in synchronism with the national grid on import control.

3.BOILER MODIFICATIONS

An existing boiler was chosen to be modified and act as a waste heat recovery boiler. The choice of Boiler 3 was made primarily because of its location allowing ease of access for turbine exhaust ductwork and also because of its capacity to accept the turbine exhaust flow volume.

The boiler was a Foster-Wheeler/John Brown balanced draught, natural circulation watertube boiler and was rated at 63.5 tonne/hour at 46.2 barg, 415°C (140,000 lb/hr, 670 psig,780°F) and was fired on HFO (Heavy Fuel Oil). The boiler is fitted with both primary and secondary superheaters and an economiser. The induced draught fan has both an electric motor drive and an auxiliary steam turbine drive. The modifications allowed the boiler MCR (Maximum Continuous Rating) to be achieved on a new Cold Air Mode (CAM)gas fired system or on Turbine Exhaust Gas (TEG) with supplementary gas firing. The specification called for the modified boiler to maintain the rated steam output and conditions, efficiency was to be equal to or better than the original at 92.3% (LCV).

The superheater surface area was reduced by removal of sixty tubes from the secondary superheater to maintain final superheat steam temperatures within acceptable limits for the outlet header and distribution system.

The economiser is a composite Welded Steel Gill tpye supplied by E.Green and Son Ltd.UK. It had a surface area of 1356 sq.m. and a gas exit temperature of 196°C (385°F) on HFO. It was extended to 2573 sq.m. to maximise the heat recovery from the turbine exhaust volume. The exit temperature is now 135°C(275°F) on TEG and 132°C (270°F) on CAM. This was achieved by an extension of the economiser on top of the original.

The original oil burning equipment was removed and the wind box modified to take the exhaust gas duct from the gas turbine. The main contractor for the boiler modifications sub-contracted the boiler burner equipment and boiler management system to Rodenhuis and Verloop b.v. Holland. The boiler management system interlinks with the Centrax gas turbine controls where necessary. The burner system is cabable of firing to 100% MCR on CAM or Supplementary firing to 100% MCR on TEG mode. Two natural gas burners are arranged one above the other on the furnace front and each is ignited by natural gas fired pilot igniter and monitored by two UV flame detectors. The two burners have a common control system and cannot be operated separately.An existing gas line suppling Boiler 1 and 2 was extended to Boiler 3 and a new gas train added. All gas supply, firing and combustion control equipment is designed and installed in accordance with British Gas Codes 17/18, IM/16 & IM/2, British Standard 5345 Part 1 (1976), and NFPA 85 B.

The contract for the boiler modifications which was awarded to Aalborg Boilers, Denmark also included the ductwork from the gas turbine attenuator to the new windbox and the ductwork to and including a gas turbine bypass stack. The ductwork was fabricated in 1% Cr.0.5% Mo. steel and stiffened externally with carbon steel flanges.

4. GAS TURBINE GENERATOR SET AND AUXILIARIES.

The gas turbine generator set is a Centrax CX571 consisting of an Allison 571-K gas turbine driving a Brush BRSDW 103/144.4 a.c generator and auxiliaries. The specific fuel consumption was quoted as 11560 kJ/kWh.(7510 Btu/hph) for the gas turbine.

The Allison 571-K industrial gas turbine is a front drive, two shaft,free turbine machine incorporating variable geometry within the compressor, air cooled blades and vanes, an annular combuster, a two stage gas producer turbine and a three stage power turbine. The engine is an aeroderivative industrial engine weighing only 807 kg (1780 lbs).

The compressor is a 13 stage axial flow with the first 5 stages having variable geometry and a compression ratio of 12:1 at ISO conditions.

The natural gas is distributed in the annular combuster by 16 fuel nozzles. Ignition is achieved by two high energy spark ignition plugs.

The two stage gas producer turbine is directly connected to the compressor rotor, it also drives engine accessories through an accessory gear train. The blades and vanes are air cooled which reduces the average nozzle metal temperature by 175°C to 250°C below the surrounding gas temperature. This section runs at 14,500 rpm .

The power turbine is a three stage uncooled free running unit. The power is transmitted via a concentric shafting system to the main gearbox. This turbine typically runs at 11,400 rpm.

The reduction gear box is an Allen ASG 29 epicyclic double helical overhung star type design and is is directly attached to the flanged shaft of the a.c generator.

The engine, gearbox generator and pumps are mounted on a common bedplate and housed within a sealed accoustic enclosure which in addition to limiting noise from the set also provides controlled ventilation and limits the area for fire and gas leak detection.

The air for the gas turbine is provided through a high velocity multistage coastal zone rated air filtration system mounted on the roof of the gas turbine house. The filter parts are of stainless steel construction for a marine environment to minimise corrosion and maintenance.

The heat rate for the engine is 11560 kJ/kWh at 5.4 MW (ISO) The engine consumes in excess of 2000 cu.m /hr of natural gas at 22 barg

The a.c. generator is a foot mounted, self-ventilating, revolving field, brushless machine with brushless exciter, and automatic voltage regulation.

The gas supply to the site is delivered at 15 barg and it was therefore necessary to install compressors to provide the gas at 22 barg for the gas turbine. Two Bellis & Morcom VL28-01N, two cylinder vee, single stage water cooled, oil free reciprocating gas compressors were installed. These are motor driven through a belt drive and mounted on a common base frame with a receiver. The receiver is necessary to provide sufficient storage to allow compressor changeover without loss of supply to the gas turbine. Each unit is 100% rated and they are arranged for automatic start of the standby unit on failure of the running unit.

The compressors are housed in a seperate building which is classified as a Zone 1 hazardous area. The compressor control panel is housed in the gas turbine house which is classified as a non-hazardous area because the hazard is contained within the gas turbine enclosure.

The gas turbine and generator control and monitoring panels are housed in the main power house controlroom. A PLC controls start, stop, crank and waterwash sequences and provides monitoring for temperature, vibration etc, necessary for the safe operation of the equipment.

5. ELECTRICAL HIGH TENSION SYSTEM.

Electrical generation is at 10kV nominal (10.6kV in practice) and is connected to the plant's main 10kV switchboard via an isolating transformer.

In Ireland the 10kV national grid is an unearthed overhead line network and before a consumer can synchronize in-house generation with the grid it is necessary to comply with a number of strict requirements, one being that equipment must have a specific impulse insulation level of 95kV. As the manufacture of such a generator was prohibitive an oil filled, double wound 10kV/10kV isolating transformer was fitted.

The plants main switchboard was modified to take the new generator circuit breaker by converting an existing standby supply breaker. The standby supply has been re-engineered. Both the main incoming supply circuit breaker and the generator circuit breaker are now controlled from the power house controlroom. Check synchronising facilities have been added which allows automatic or manual synchronising in either direction from the generator control panel. The system incorporates full electrical protection on switchgear, generator and auxiliaries.

A secure 415V supply is provided for the generator set auxiliaries from an LV switchboard which is fed from a 10kV/415V transformer from the main 10kV switchboard to which the generator is connected. A new 800kVA black start diesel generator was installed as part of the project to provide an alternative 415V supply to the LV switchboard in the event of loss of supply from the main 10kV switchboard. This diesel generator has capacity in excess of the requirements stated above and is used for emergency supply for other power house equipment mainly Boiler 3. It is skid mounted and acoustically enclosed. The black start generator supply and the mains supply to the LV switchboard are mechanically and electrically interlocked. An auxiliary transformer gives further security to the 415V supply to the gas turbine generator auxiliaries and is fitted with auto changeover facilities. (see appendix 1.)

6. PLANT OPERATION

The manufacturing process on site demands a high level of security and it is therefore Pfizer policy to run the Utilities Plant in a manner which gives that security. The plant is run on a 24 hour day, 7 day week basis with a 2 day break at Christmas and a very short summer shutdown. The steam demand is capable of being supplied by one of three boilers but two are normally on line. The normal operating mode would be Boiler 1 or 2 on a base load and Boiler 3 responding to the fluctuating demand of the plant and taking maximum advantage of the oxygen rich hot turbine exhaust gas. The turbine generator set is run in parallel with the national grid on import control as export for more than 5 seconds is not permitted and will cause the generator circuit breaker to trip.

This is the normal mode of operation necessary to satisfy the needs of the plant and at the same time maximise the savings from the CHP system.

The average electrical demand for the site was 4.68 MW with peak demands of 6.2 MW approximately. Therefore, for the majority of the time the plant will operate within the capacity of the generator set. The load on the gas turbine generator is controlled by the import controller up to peak temperature of 803 C at which point the power demand exceeds the rating of the generator set the controller will automatically import the required demand above the generator set capacity from the national grid. Import control is necessary because it is prohibited to export to the grid in Ireland. The system can be run in parallel with the grid or independantly in "island mode" . In parallel with the grid a minimum import of between 50kW and 100kW is necessary to maintain the set in synchronous with the grid.

In "island mode " governor control is on the power turbine while in parallel operation with the grid governor control is on the gas producer turbine and is a smoothed signal. This was not the original design mode of governing but was a modification necessary to allow the generator set to run in parallel with the frequency and voltage fluctuations on the grid. The effect of these fluctuations was realised only when some operating experience was gained with the system and it was realised that control modifications would be necessary in the early weeks of operation.

The original electrical interlocks and controls called for the generator circuit breaker to open in the event of a problem with the national grid such as voltage dips experienced during an electrical storm, but for the national grid circuit breaker to remain closed. This was because of the relatively low inertia of the gas turbine. This type of undervoltage problem usually results in loss of much of the electrical plant on site due to undervoltage tripping and/or contactor drop out. However, the system has been modified recently to trip the grid circuit breaker and to shed about 25% of the plant load and allow the gas turbine generator set to remain on line thereby hopefully avoiding loss of supply to the production areas of the plant. The emergency diesel generator supply available on site should cater for the plant affected by the load shed. This system still awaits a live test.

7. PERFORMANCE DATA.

7.1 Waste Heat Recovery Boiler.

Formal performance tests were carried out on the CHP system on 28th and 29th July 1988. The delay was due to a number of gas turbine related problems.

The boiler performance test was conducted in accordance with BS 2885: 1074 - "Acceptance tests on stationary steam generators of the power station type". Readings were taken to enable boiler efficiency to be calculated by both Method A (direct method) and Method B (losses method).

Method B results were considered more reliable since they are less dependant on flow measurement where greatest inaccuracies occur. The Method B results are used for comparison with anticipated results and are shown in Table 1.

Since greatest discrepency with anticipated results occurs at 25% MCR on CAM and is only -1.8%, the results were considered acceptable.

7.2 Gas Turbine Generator

The gas turbine performance tests were conducted in accordance with BS 3135:1975 - "Specification for Gas Turbines: Acceptance Test".

Guaranteed performance figures and actual test results are shown in Table 2. Specific fuel consumption was found to be 0.1% above the maximum limit of the guarantee value of 105.26% . This was accepted bearing in mind the complexity of the test and influence of change in ambient temperature and pressure on GT performance, although with reservations.

7.3 Overall Performance.

The overall CHP performance test results were within acceptable limits of the predicted efficiencies calculated from the guarantee values submitted for the boiler and the gas turbine. These results are shown in Tables 3.

8. ECONOMICS.

The economic results of the CHP system are directly affected by the price of natural gas and by the price of imported electricity from the national grid. Both fuel and electricity fell soon after the project was commissioned and have continued to remain below those on which the project payback was originally estimated. The savings however continue to give a satisfactory return on the investment.

Acknowledgements to

Ewbank Preece
Centrax
Aalborg Boiler
Rodenhuis & Verloop

TABLE 1

COMPARISON OF BOILER EFFICIENCY RESULTS WITH ANTICIPATED VALUES

Anticipated values (% efficiency based on LCV) : -

Boiler % MCR	Gas turbine load (kW) 5400	2500	Cold Air Mode
100	93.3	93.8	93.8
50	88.3	90.7	93.8
25	76.0	81.5	92.3

Test results : -

Boiler % MCR (Nominal)	Method	Gas turbine load (kW) 5300	2500	Cold Air Mode
100	A	92.4	93.3	97.9
	B	93.2	94.3	93.6
50	A	86.0	88.9	90.9
	B	90.6	91.6	92.6
25	A	83.7	83.7	96.3
	B	88.1	86.7	90.6

TABLE 2

GAS TURBINE PERFORMANCE

Guaranteed Performance

Nominal Power Output	: 5370 kw
Nominal Specific Fuel Consumption	: 11560 kJ/kWhr
Conditions :	
Ambient temperature	: 15° C
Atmospheric Pressure	: 1013 mbar
Intake pressure drop	: 100 mm H2O
Exhaust back pressure	: 380 mm H2O
Power turbine speed	: 11413 rev/min
Fuel	: Natural gas

Including gearbox and generator losses with generator at 0.8 to unity
Limits : - Power - 95 %, SFC + 105.26 %.

Typical Site Conditions

(@ 1515 on 28th July 1988)

Ambient temperature	: 18°C
Atmospheric pressure	: 1004.75 mbar
Intake pressure drop	: 84 mm H_2O
Exhaust back pressure	: - 38 mm H_2O

Test data corrected to ISO conditions

Power	: 5491 kWe
Specific fuel consumption	: 12076 kJ/kWh

Centrax fax dated 18th November 1988

Specific fuel consumption at 5370 kW power output (by i

= 12180 kJ/kWh
= 105.36 % of guaranteed

TABLE 3a

OVERALL CHP EFFICIENCIES

1. Guaranteed values at gas turbine generator and boiler full load conditions.

A	kW	41,638
B	kW	17,244
C	kW	48,502
D	kW	5,370

Efficiency $\left(\frac{C+D}{A+B} \times 100 \right)$ % = 91.50

Tolerances + 0.86 %
- 5.33 %

Therefore acceptable efficiency range is 86.6 % to 92.3 %

2. Test results, gas turbine generator at full-load.

Boiler load (% MCR)		25	50	75	100
A	kW	4,072	15,635	29,105	41,522
B	kW	16,362	18,271	18,108	18,139
C	kW	12,191	22,970	36,589	47,097
D	kW	5,360	5,360	5,230	5,310
Effic. $\left(\frac{C+D}{A+B} \times 100\right)$ %		85.88	83.55	88.58	87.84

3. Test results, Gas turbine generator at half-load.

Boiler Load (% MCR)		25	50	75	100
A	kW	6,950	19,102	32,961	40,422
B	kW	10,680	11,048	11,019	10,785
C	kW	11,785	23,877	36,416	44,677
D	kW	2,520	2,550	2,580	2,486
Effic. $\left(\frac{C+D}{A+B} \times 100\right)$%		81.14	87.65	88.67	92.10

TABLE 3b

TEG %	100				50			
BOILER %	25	50	75	100	100	75	50	25
1 Mean ambient temp at g.t. air intake C	17.77	18.17	19.70	20.17	18.90	17.85	17.50	16.35
2 (D) G.T. electrical output kW	5,360.00	5,360.00	5,230.00	5,310.00	2,486.00	2,580.00	2,550.00	2,520.00
3 Boiler steam flow kg/s	4.71	8.62	13.40	17.14	16.25	13.31	8.91	4.55
4 Boiler steam net enthalpy kJ/kg	2,538.30	2,664.73	2,730.50	2,747.60	2,745.96	2,736.00	2,679.80	2,590.20
5 (C) Boiler output (3x4) kW	12,191.00	22,969.97	36,588.76	47,097.00	44,676.69	36,416.16	23,877.00	11,785.40
6 Total CHP output (2 + 5) kW	17,551.00	28,329.97	41,818.76	52,407.00	47,162.69	38,996.16	26,427.00	14,305.40
7 Gas fuel to g.t. kg/s	0.331	0.3693	0.366	0.3667	0.2180	0.2227	0.2233	0.2159
8 (B) Fuel energy to g.t. (7xC.V.) kW	16,362.00	18,271.20	17,107.90	18,139.00	10,785.10	11,019.20	11,047.48	10,680.17
9 Gas fuel to boiler kg/s	0.079	0.3122	0.584	0.8338	0.8126	0.6625	0.3827	0.1370
10 Fuel energy to boiler (9xC.V.) kW	3,889.70	15,445.76	28,891.70	41,251.00	40,201.00	32,775.20	18,930.47	6,781.06
11 Aux. power cons boiler kW	182.70	189.20	213.10	271.20	221.00	186.00	171.90	168.60
12 (A) Total energy to boiler (10+11) kW	4,072.40	15,634.96	29,104.80	41,522.20	40,422.00	32,961.20	19,102.40	6,949.70
13 Total energy to CHP (8+12) kW	20,434.40	33,906.20	47,212.70	59,661.20	51,207.10	43,980.40	30,149.90	17,629.80
14 CHP efficiency (6/13 x 100) %	85.88	83.55	88.58	87.84	92.10	88.67	87.65	81.14

Items A, B, C, D, refer to Section 3.0 of the report.

APPENDIX I

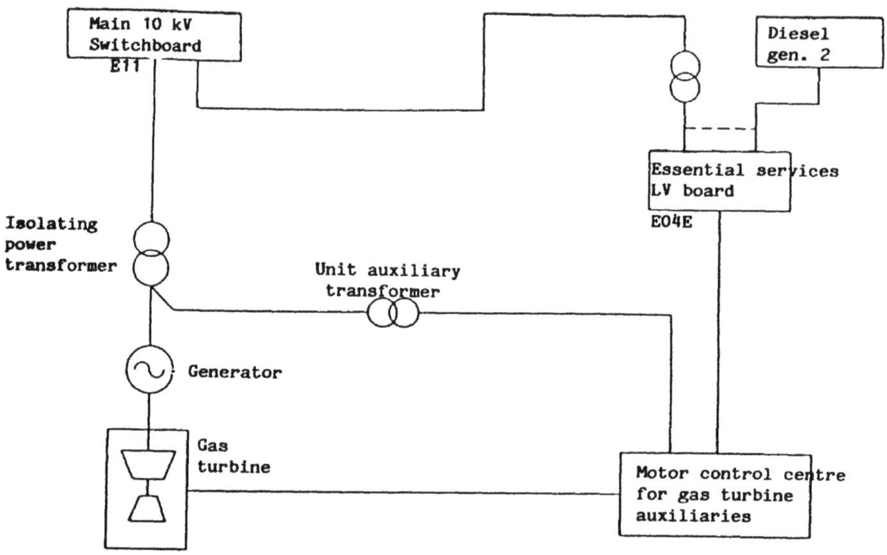

Diagram of gas turbine auxiliaries power supply

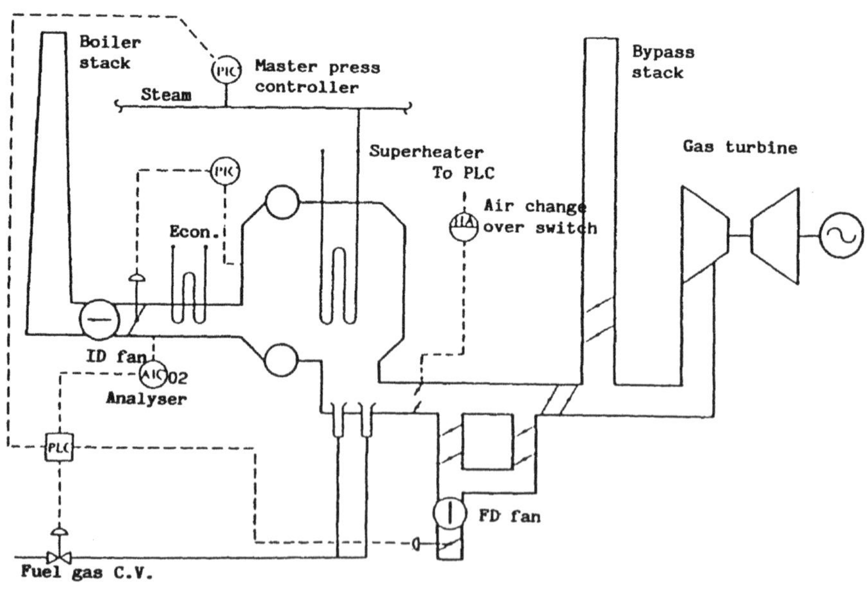

Diagram of conversion of Boiler 3

HUNDRED THOUSAND HOURS BASELOAD COGENERATION WITH THE IM-5000

E. HOLLROTTER

Dow Stade GmbH
Germany

SUMMARY

Dow Stade process energy (power and steam) is supplied by a gas turbine driven power plant with five gas turbines and six steam turbines with four auxiliary package boilers. Due to increasing natural gas prices and improvements in the process plant there has been an inbalance in the supply and demand of energy and high utility costs. The scales have been brought back balance by replacing three of the heavy duty FIAT-TG20AA gas turbines with three IM-5000 gas turbines (aircraft derivatives). Hydrogen supplied by the process plants is also burnt in the gas turbines instead of burning in the boilers. Now, after 100,000 operating hours at Dow, the utmost high thermal efficiency has been maintained over the years, the enrergy savings have been higher than predicted but achieved work availability has been lower than the targeted 95 %.

RESUMEN

La energía de proceso (vapor y electricidad) en Dow Stade se suministraba mediante una central de turbinas de gas con cinco turbinas de gas y seis de vapor junto con cuatro calderas auxiliares. Debido al incremento de los precios del gas natural y la mejora de los procesos, se presentó un desequilibrio entre oferta y demanda de energía y altas facturas energéticas. La situación ha vuelto al equilibrio al sustituir tres de las grandes turbinas FIAT-TG20AA por tres turbinas a gas IM-5000 (derivadas de la aviación). El hidrógeno generado en los procesos se quema también en las turbinas en vez de en las calderas. En la actualidad, después de 100.000 horas de funcionamiento en DOW, la más alta eficiencia térmica se ha mantenido en estos años y los ahorros de energía han sido superiores a los previstos, sin embargo, la disponibilidad ha sido inferior de la prevista del 95 %.

HUNDRED THOUSAND HOURS BASELOAD COGENERATION WITH THE IM-5000

E. HOLLROTTER

Dow Stade GmbH
P.B. 1120
D-2160 Stade

Introduction

Dow Stade is situated in the very north of Germany direct at the Elbe river and close to Hamburg (see Fig. 1). The production for the commodities is based on the production and use of chlorine for conversion. Salt stocks 25 km from the plants are used to supply the site with brine. The The electrolysis of brine is highly power consuming and approximately 80 % of the electricity is used in the chlorine cells. Major steam users are the caustic evaporator, the epi-chlorhydrine plant and the propylene oxide plant (Fig. 2). Start up of the plants has been in 1972 and the plants of the first site extension program came on line end of the seventies. With these projects the burning of hydrogen ex chlorine cells became operating standard. This major energy saving project opened in line with conversion energy savings in the propylene oxide and caustic plant the gap between demand and supply.

Operating all gas turbines and using all hydrogen from the chlorine plants would have resulted in an excess of approximately 80 MW/h thermal energy. This reduced the cogeneration capability with secondary result of higher amount of purchased power and increased energy cost for the total site.

In addition to this, the continuous escalating gas prices combined with the possible savings in five plants opened for Dow an opportuinity to solve the problem by increased cogeneration with the most efficient gas turbine available and applying all the conversion energy savings at the plants (Fig. 3). Projections also showed that the power demand in the future is increasing faster than the process heat demand. The ratio of thermal energy to power would change more and more to the power site (Fig. 4). Two projects have been authorized, consisting of one important condition. The major part of the first package was the exchange of one TG 20 AA by one IM-5000 and the optimation of the heat recovery in the power plant and usage in the plants (see Picture 1, Table 1).

Project

The selection of the LM-5000 for baseload cogeneration was based on the following key points: Outstanding high thermal efficiency of the LM-5000, power output in a range that fits optimally the cogeneration requirements of the Stade site, the experience heritage of the LM-5000 from CF6-50 with approximately 30 million operating hours and the LM 2500 with approximately 3 million operating hours and last not least a parts communality of approximately 70 % (Fig. 5). The total LM-5000 operating experience at the time Dow decided to install the LM-5000 was less than 30 000 operating hours and most for peak power production and not for base load cogeneration. The comparison of the characteristic operating datas for both types of gas turbines show the big step towards highest efficiency but also for best performance monitoring systems for this advanced technology (Fig. 6).

The installation of the two types of the gasturbines is totally different. The FIAT gas turbines are surface insulated and the LM-5000s are operating in an enclosure. This enclosure covers both, the noise- and the heat insulation. For surface heat removal the enclosure is cooled by approximately 30 000 m3/h air, which is compressed from the filterhouse into the enclosure (Picture 2, Picture 3).

Calculation for the existing heat recovery system with the operational datas of the LM-5000 showed, that there is enough margin for the superheated steam temperature requirements and also room for increasing the economizer section (Fig. 7). There have been five plants involved in the Stade site cogeneration improvement. The extension of the heat recovery units is shown in picture 4. In one of them was an opportunity to replace low pressure steam by producing flash steam with preheated bottom effluent in the heat recovery section (Fig. 8). The energy losses of the cycle have been reduced from 31.6 % down to 13.4 % of one gas turbine system, while on the other hand the electric power output of the cogeneration unit increased from 31.6 % to 40 % (Fig. 9).

Operating Experience

Now, Dow Stade has accumulated 100 000 hours operating baseload cogeneration with LM-5000s (Picture 5, Table 2). A lot of operation experience has been collected and improvements been made. We experienced in the beginning that operation of aircraft engines with datas of maximum 10 hours of operation are different to those on continuous full load operation with air density at sea level. There had been a sophisticated online monitoring system installed which is calculating every minute performance datas on a corrected base. There are seven base performance datas which are important for reliable operation. For these performance datas there are limits set which allow maximum and absolute deviation from the original condition. These are: Overall compressor efficiency , variable stator vane deviation from nominal value, corrected power output, corrected fuel flow, corrected pressure ratio of the high pressure compressor, corrected pressure ratio of low pressure turbine inlet ÷ atmosphere, corrected temperature rise over the low pressure compressor and mechanical performance datas. By trending of these performance datas it was found in the beginning that the high pressure compressor is extremely sensitive on contamination, resulting in efficiency drop to the allowed limit within four weeks or less.

For reliable baseload cogeneration uniterrupted operation over two or three months is required. Therefore the concentation of work over the years was to increase the efficiency of the air filtration system to the utmost achievable. Now, we are running the gasturbines with no compressor efficiency degradation for over three months. The inlet filtration system is also equipped with evaporative cooling and anti-icing (Fig. 10)

Improved monitoring of mechanical performance datas of the oil system, vibrations of the total turbine set with spectrum analyzer, trending datas with statistical analysis of the combustion area also improved the operation that even small deviations from nominal can be detected in an early state. Problem areas hade been in the beginning the high presssure compressor with the exact and durable control of the variable stator vanes and shifted with longer operation to the hot parts: Most of the problems have been solved in cooperation with IHI and GE (Picture 6). As expected from the beginning, the heavy duty power turbine was trouble free. Now houndred thousand operating hours are achieved and also with the time the knowledge how to operate aeroderivative gas turbines in an environment, where reliability counts first.

Conclusion

Taking all the experience and savings with the IM-5000 in account, I can summarize that the decision of installing LM-5000 with the heavy duty Japanese power turbine was right. The total LM-5000 fleet in the world increased to 18 units in total, where 13 of them are cogeneration units. Except the 4 units running and the Dow units, all of the other LM 5000 / IM 5000 operators are in the United States, showing that those operators are going faster in innovative products. Dow Stade has now accumulated 100 000 fired hours, representing close to one third of the total 331,000 accumulated fired hours. The experience is increasing with the number of the running engines and problems are decreasing with the increased knowledge how to operate aeroderivative gas turbines in industrial service. This knowledge will help us to operate the turbines of the next generation with potentially even higher firing temperature, higher compression ratio, might be in combination with pressurized fluidized bed combustors or fuel cells (Table 3).

References

(1) LM 5000 for Cogeneration and Power Generation printed September 1989

Fig. 1: Location of Dow Stade GmbH

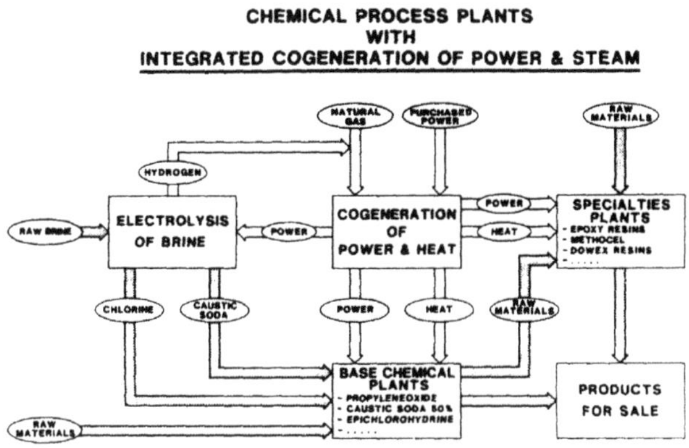

Fig. 2: Integration of Cogeneration in Stade

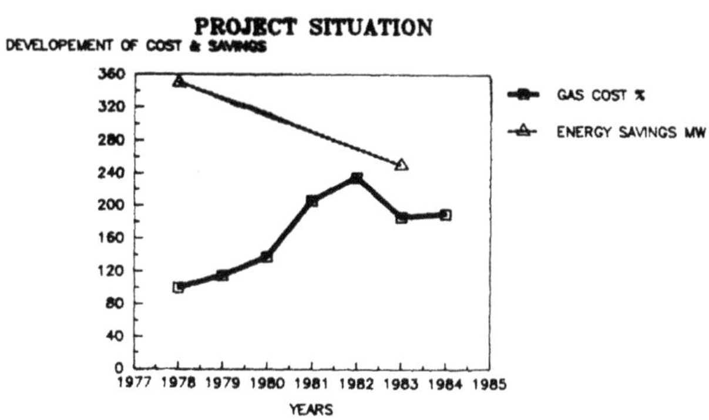

Fig. 3: Situation for Dow Stade at Project Start

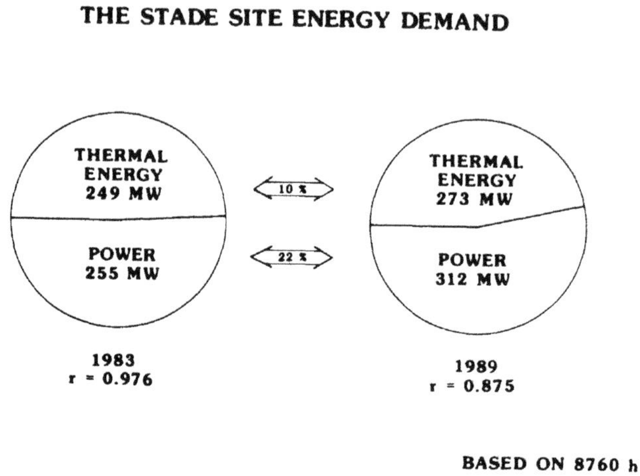

Fig. 4: Comparison of 1983 and 1989 Operation

Picture 1: Typical Heat Recovery Section at Dow Stade before Start of the Project

GASTURBINE COMPARISON

		FIAT TG20AA	IM 5000
SHAFTPOWER	[MW]	25	35.1
HRU-STEAM	[t/h]	70	56
ELECTRICAL EFFICIENCY (LHV)	[%]	24	35.3
FIRING TEMPERATURE	[°C]	913	1160
EXHAUST TEMPERATURE	[°C]	450	426
EXHAUST FLOW	[kg/s]	155	130
SHAFT SPEED	[rpm]	4860	3720/10380

Table 1

Figure 5

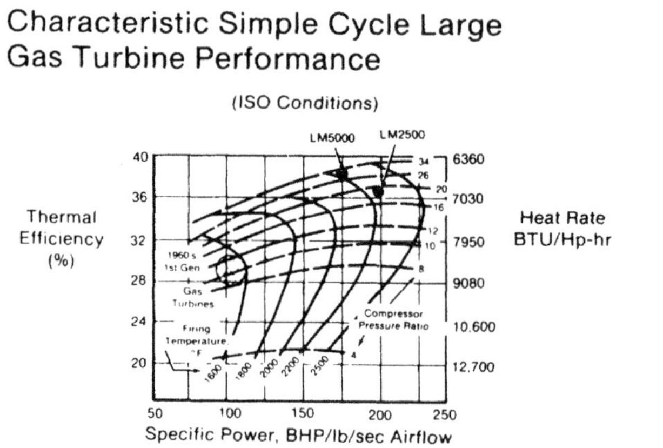

Figure 6 (1)

Picture 2: Industrial Gas Turbine

Picture 3: IM-5000

PINCH POINT DIAGRAM OF HEAT RECOVERY UNIT WITH IM 5000 COGENERATION CYCLE

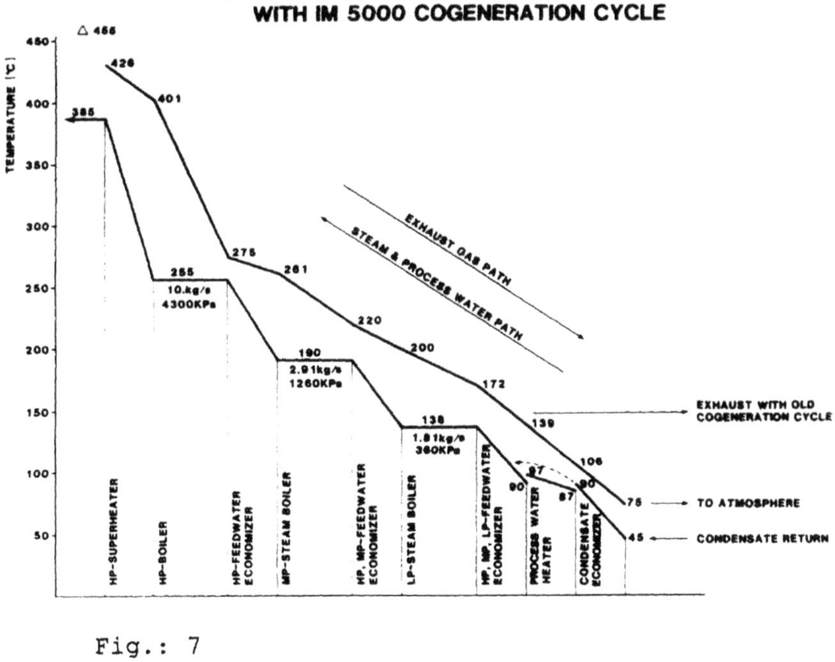

Fig.: 7

Picture 4: Typical Heat Recovery Unit after Modification

PROCESS DIAGRAM
COMBINED CYCLE COGENERATION WITH IM 5000 GT

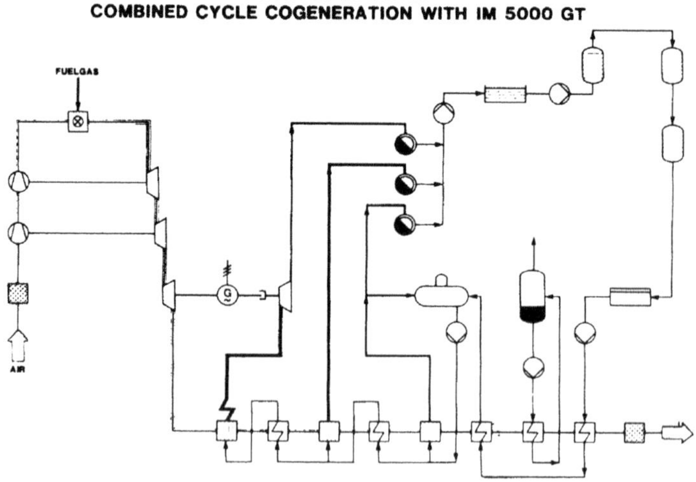

Fig. 8

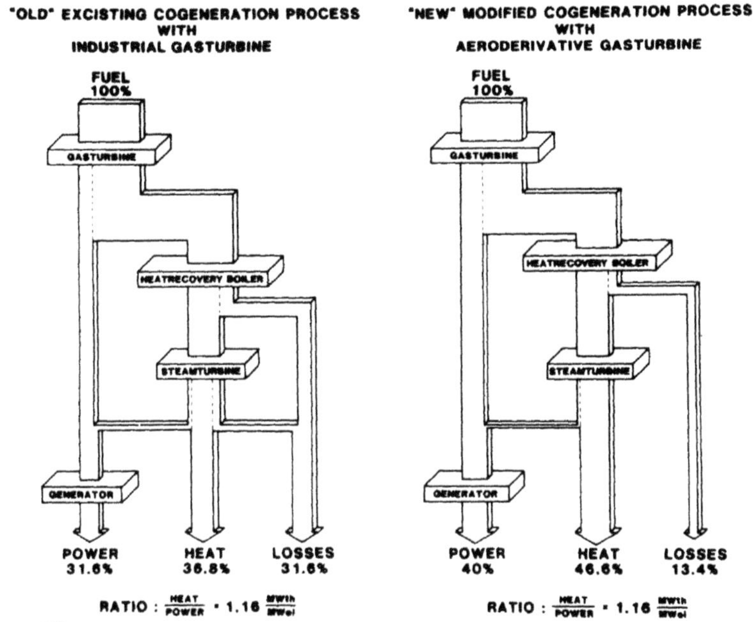

Fig. 9:

Picture 5: Turbine Hall No. 1 with 4 Gas Turbines

IM - 5000 Operating Status *

Turbine	EGT-4	EGT-1	EGT-2
Start up	Feb 84	May 85	May 86
Operating hours	42,316	31,649	24,683
Starts	258	131	88
Power generated Mwh	1,679,350	1,067,993	901,113

* as of September 1989

Table 2

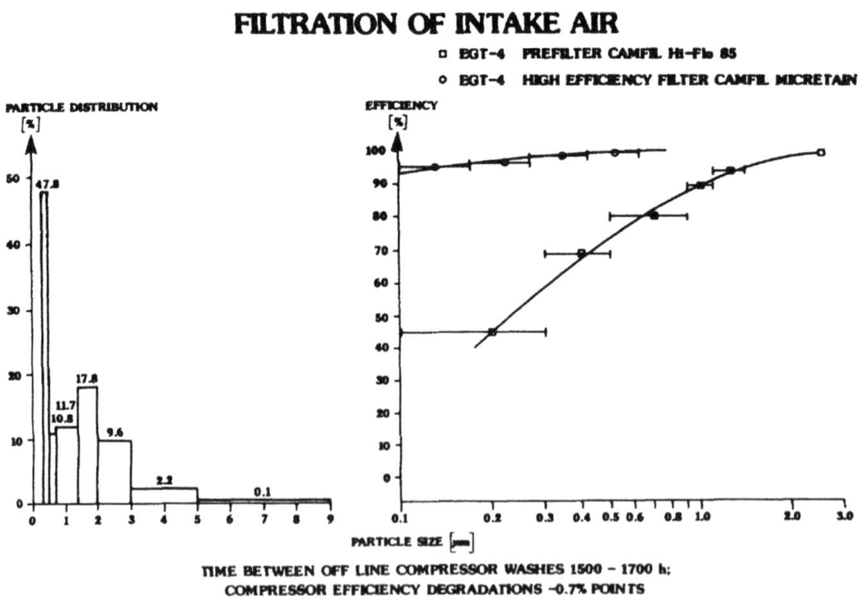

Fig. 10:

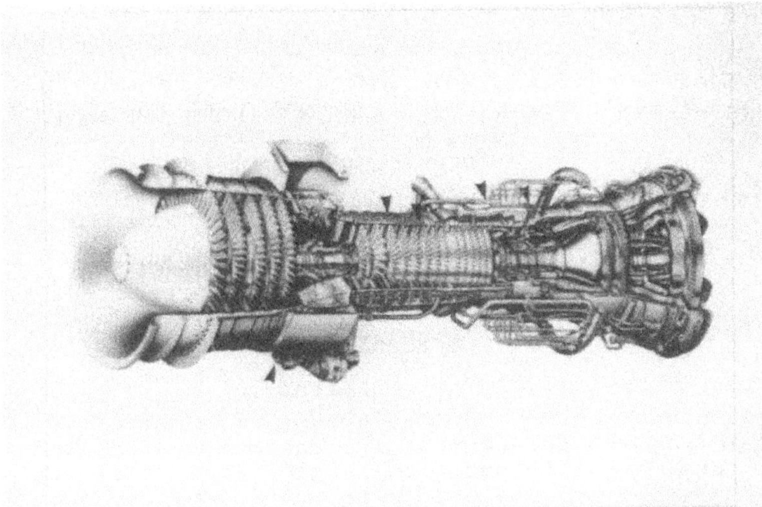

General Electric LM5000 Gas Generator

Picture 6 : (1) Problem Areas

LM5000 Installations

Installation	Date
Meidensha Electric Numazu, Japan	8/77
Bangladesh Barges Khulna, Bangladesh (2 units) Chitagong, Bangladesh (2 units)	 4/80 7/86
Simpson Paper Co. - Anderson, California - Pomona, California - Ripon, California	 5/83 12/85 4/88
Dow Chemical Co. Stade, W. Germany (3 units)	2/84 2/85 2/86
Catalyst Energy Dev. Corp. Oildale, California	12/84
Container Corp. of America Vernon, California	4/86
Corona Cogeneration Partners Corona, California	5/88
Reedy Creek Utilities Disney World - Orlando, Florida	9/88
Univ. of Northern Colorado Greeley, Colorado (2 units)	4/88
Greenleaf Power Yuba City, California (2 units)	12/88 11/89
New Brunswick Power Grand Manon Island, New Brunswick	6/89
Energy Factors N. Island, San Diego, California	6/89
Carson Energy/Ice Haus Carson City, California	7/89
O'Brien Energy/Merchants Refrigeration Salinas, California	9/89
Proctor & Gamble Co. Oxnard, California	10/89
Power Systems Engineering Bakersfield, California (2 units)	12/89
Tropicana Products Bradenton, Florida	12/89
CNG Energy Lakewood, New Jersey (4 units)	5-7/90
Coastal Power/Nestles Fulton, New York	7/90
Agra Power Salinas, California (2 units)	12/90

Data as of March 1989

Table 3

HUNDESTED DECENTRALIZED HEAT AND POWER PLANT

PER LOETH

Elselskabet EFFO
Denmark

SUMMARY

Based on experience of other CHP-systems, the actual CHP-based heat production has been calculated to cover approximately 40-50 per cent of the maximum heat demand (90 % of the total heat consumption) while the remaining heat demand is covered by a peak load production based on ordinary (gas or oil) heat boilers. The plant consists of the following systems: 1 A medium speed 4-stroke dual-fuel engine with direct heat recovery for district heating. 2 A heat pump for utilization of low temp. cooling. 3 A heat storage system capable of storing heat for delivery within 24 hours. 4 A boost fired boiler for peak load heat capacity. This system has reached high total efficiency (96,5 %), and can change-over from gas to oil during gas grid peak load periods. It has also achieved concentration of electricity production at low heat demand by storing heat during the day for subsequent delivery at night. The CHP plant is expected to be 2.34 Mw-e and 3.92 Mw-heat.

RESUMEN

Basado en la experiencia de otros sistemas de producción combinada de calor y electricidad (CHP), el sistema actual ha sido calculado para suministrar aproximadamente entre el 40 % y el 50 % de la demanda máxima de calor (el 90 % del consumo total de calor), el resto de la demanda se cubre con una producción de puntas basada en calderas convencionales (de gas o productos petrolíferos). La planta posee los siguientes sistemas: 1 Un motor de 4 tiempos, de velocidad media multicombustible con recuperación directa de calor para la calefacción urbana. 2 Una bomba de calor para la utilización del calor de refrigeración de baja temperatura. 3 Un almacenamiento capaz de almacenar calor para suministrarlo en las 24 horas siguientes. 4 Una caldera de choque para suministrar calor en punta. Este sistema ha alcanzado una alta eficiencia (96,5 %), puede cambiar de gas a petróleo en las horas punta de la red de gas, concentrar la producción de electricidad en los valles del calor y almacenar el calor sobrante durante el día para suministrarlo por las noches. La planta se espera que produzca 2,34 Mw-e y 3,92 Mw térmicos.

HUNDESTED DECENTRALIZED HEAT AND POWER PLANT.

Loeth, Per
elselskabet EFFO

Heat demand.

The size of the decentralized heat and power plant at Hundested has been determined by the expected heat consumption.

The heat consumption for heating and hot water is not only subject to seasonal variations but it also varies within 24 hours. The heat consumption in a district heating system is usually described by a "load profile". This load profile expresses the number of annual hours in which the actual demand exceeds a certain value.

Based on experience from other CHP-systems, the actual CHP-based heat production has been calculated to cover approximately 40-50 per cent of the maximum heat demand, while the remaining heat demand is covered by a peak load production based on ordinary heat boilers (gas or oil boilers). This distribution has proven to produce the most economical system as a coverage of approximately 50 per cent of the maximum heat demand means that approximately 90 per cent of the total heat consumption is covered by heat and power production. If the remaining part of the heat demand was to be covered by CHP-produced heat this would result in too low energy production and too low profits from the investments (this is of cource due to the fact that it is cheaper to install ordinary boilers than to establish production plants for heat and power).

The dimensioning of the heat and power plant is based on the "Draft Heat Plan for the Municipality of Hundested, November 1984".

The economic calculations made in connection with the heat and power plant assumes that the rate of connection to the heat and power network is increased during the construction period. The present rate of connection is approximately 50%. This is expected to increase to 75 per cent with a linear increment during a 10 year period.

The district Heating Network.

It is necessary to know the network losses in order to determine the heat demand on the basis of the information available from Hundested heating plant, as there are no exact energy measurements. In the heat prognosis of the district heating system an improvement of the operating conditions of the district heating system is assumed. In addition to renovations of pipeline sections of poor condition, this includes removing by-passes in connection with introducing calorimetres and thus lowering the return temperature which reduces the heat loss in the pipelines.

When measuring energy consumption the forward temperature can be lowered which will reduce the heat loss. This will improve the economy for the consumers of district heating and increase the efficiency at the CHP-plant when the return temperature is lowered to 50 degree C.

The Siting of the CHP plant.

It has been considered to locate the CHP plant at the existing district heating plant. However, the conclusion from these considerations is that due to environmental aspects (noise) it is considered to be either impossible or to involve heavy expenditure to observe the demand of not more than 35 dB (A) in proberty boundary. The plant should therefore be placed in areas where a noice level of 40 dB (A) - which is the expected level from the plant - is acceptable, for example the industrial area. This siting involves establishing a district heating pipeline from the heat and power plant to either the old district heating plant or to another point of the district heating network from which the heat can be distributed in the existing district heating network.

Security of Supply/Supply Principles.

In addition to the actual heat and power unit with a capacity of approximately 4 MW heat, a boost fired boiler with a capacity of approximately 5 MW heat has been added to exhaust gas system. This boost fired boiler is employed when the heat and power unit is unable to produce the necessary amount of heat but it is also fitted with a fresh-air blower which enables it to work independent of the heat and power unit. The heat and power plant is also fitted with a separate gas boiler with a capacity of approximately 5 MW heat.

Thus the plant comprises 3 units of approximately 5 MW heat each which can produce concurrently or independently. As the dimension basis sets a maximum heat demand of approximately 10 MW heat, this can be covered by only 2 units except from a few hours of the year.

Heat storage.

The heat and power plant is connected to an heat accumulation tank. The main function of this heat storage is to secure a maximum electricity production during daytime of summer. Thus the plant is secured maximum payment for the power. In the case of maximum electricity production in summer the district heating network is unable to dispose of the produced amount of heat from the heat and power unit. This "surplus production" is therefore accumulated in the heat storage. When the power plant is stopped during night the heat storage is emptied to meet the heat demand in the district heat system. In addition to the higher power payment it is secured that it is not necessary to operate with part load on the power plant unit. Part load usually results in a reduced efficiency of the machine.

Technical Description. Prime Mover.

As prime mover for the CHP-plant has been chosen a dual-fuel 4 stroke gas engine, which offers the following advantages:

a. High mechanical efficiency.
b. Possiblity of change to oil operation during gas peak load periods. The capacity and operating costs of the gas-grid will thereby be reduced.
c. Exhaust gas suitable for boost firing, enabling additional heat production during peak load periods with good efficiency.
d. Gas pressure demand of app. 3 bar-a which is lower than expected minimum supply pressure of app.6bar-a.

Engine Type.

Since there is no danish dual-fuel engines available at the moment, a conversion of the MAN-B&W Holeby L/V-28-32 diesel engine for dual fuel operation on danish gas has been chosen. This engine type can be delivered with 6, 8, 9, 12, 16 og 18 cylinders with a mechanical output of 0.9 - 2.7 MW (Operating on danish natual gas).

Depending on the degree of heat recovery this range corresponds to a thermal output of 1.2. - 4.0 MJ/s which is suitable for smaller district heating systems with 300 - 2000 heat consummers. The engine operates at 750 rmp which is sufficiently low for base load applications. For the district heating system in Hundested a 18 cyl. engine is chosen with a nominel mechanical output of 2.65 MW corresponding to 2,53 MW at generator terminals. The mechanical efficiency will be app. 41% and the electric efficiency 39%.

When operating on natual gas as primary fuel the engine will use app. 8% of the supplied energy as diesel oil (and the rest as natural gas. The engine will be equipped with a complete diesel system, allowing full load operation on diesel fuel if necessary.

Heat Recovery System.

The heat recovery system will be seperated in two systems:

a.

A high temperature system with direct exchange of heat to the district heating system. This system includes the lubrication oil cooler, and 1 stage of the charge air cooler, which will be designed to allow a relatively high cooling temperature. The district heating water will be used directly for jacket cooling. Finally the system will include 1. stage of an exhaust gas exchanger where the exhaust gas is cooled to app. 80 deg. celcius. At nominel load app. 3.25 MJ/s heat is recovered in the high temperature system. Se attached high temperature cooling diagram.

b.

A low temperature system which supply heat for the cold side of a heat pump. This system includes the 2. stage of the charge air cooler, the generator cooler and 2. stage of exhaust gas exchanger, where the exhaust gas is cooled to app. 45 deg. celcius. At nominel load app. 0.46 MJ/s low temp. heat vill be available for the heat pump at a temp. level of App. 20 deg. celcius. Radiator coolers are installed in order to be able to operate the engine independently of the heat pump. See attached low temperature cooling diagram.

Heat pump.

In order to raise the low temp. heat to district heating level of app. 60 deg.celcius, a standard piston type heat pump unit is included. At nominel load the heat pump will require 0.19 MW electric power and will deliver 0.65 MJ/s heat to the district heating system.

Net CHP-performance.

At nominel load the plant has following net performance:

Supplied primary energi(92% gas, 8% oil):6.46MJ/s-100%
Net electricity (2.52 MW gen.-0,19 MW HP):2.33MW -36,1%
Net dist. heat (3.25 MJ/s+0,65 MJ/s HP):3.90MJ/s-60,4%

Total usefull energy :6.23MJ/s-96,5%

The remaining 3.5% of the supplied energy will be lost as radiation (2%) and stack (1.5%). Se attached Sankey diagram.

Boilers.

In order to meet the heat demand above the CHP - capacity, with a good efficiency, a boost fired boiler is included. The air excess ratio of the gas engine is app. 2.0 This allows to boost the heat capacity with app. 5 MJ/s by injection of additional gas in a boost fired boiler. The boiler will be of traditional design but a modified inlet section.

The boiler is connected in serie with the exhaust gas heat exchangers and a by-pass allows independent operation of the engine during service of the boiler.

The boiler will be equipped with a supplementary air supply, allowing the boiler to operate independent of the engine. The efficiency of the boost fired boiler will be app. 98%.

See attached exhaust gas diagram.

Heat Storage.

The minimum heat demand at Hundested is app. 1 MJ/s. During periods of minimum demand the engine will be operated at near full load during the day, and the surplus heat stored in a 600-700 cubic meter heat storage for subsequent delivery during the night.

The heat storage will act as expansion unit of the district heating system. This will ensure a stable pressure as required in order to use direct jacket cooling.

Electrical system.

The engine will be connected to the public grid through a 10 kV synchronius three-phase generator.

Electric power for the heat pump and other auxillaries will be supplied by the 380 V grid.

Control system.

The CHP plants control system will be based on a microprocessor system which will constantly monitor all relevant parameters.

The control system will be able to operate the plant unmanned for limited periods (nights).

The control system will contain optional strategies for heat demand and heat storage controlled operations.

Further more it will be possible to preprogramme a change to oil operation, when peak demand of the gas grid is expected.

Economy.

General assumptions for the economy evaluation.

Conditions and assumptions in the evaluation of operation economy are briefly described below.

Average year.

The operating economy of the CHP-plant incorporates an average year. There is a forecast for the need of district heating for 1986, 1991 and 1996.

It is assumed that year 1986 represents 1986 2 heating seasons (88/89, 89/90), 1991 represents 4 heating seasons (90/91-93/94) and 1996 represents 14 seasons (94/95-) and they are weighted with a present value factor (5% annual interest) and an average year is obtained in the 20 years calculation period with following weightings: 1986 x 0,15 + 1991 x 0,25 + 1996 x 0,60.

Heat consumption.

When calculating an average heat production as above, a 75% coverage is estimated in 1996 in district heating areas.

Connection to the network is estimated to be steady up to 1996. The district heating network is expected to be transformed to return temperatures of about 50 degrees C. to be able to reach optimum recovery of heat.

Prices of gas (HNG data sheet Jan 87.)

Calorific value 39,6 MJ/nm3

Tariffs for district heating (excl. VAT):

Annual consumption	price
< 800.000 nm3	288,7 øre/nm3 = 26,2 øre/kWh
< 5.000.000 nm3	282,9 øre/nm3 = 25,7 øre/kWh
< 15.000.000 nm3	277,0 øre/nm3 = 25,2 øre/kWh

Tarif for electricity.

There is a tax of 202 øre/nm3 corresponding to 18,35 øre/kWh for the gas used to produce electricity. The part needed for production of electricity is calculated on the basis of the total production: produced electricity(prod. el. + prod. heat).

Prices of oil (Gas oil. ref. HNG data sheet Jan. 87.

Calorific value 35,6 MJ/litre
Normal price (excl. VAT): 3139 DKK/m3 = 31,7 øre/kWh.

Tariff for electricity:

For the oil needed to produce electricity the energy tax is deducted (1850 DKK/m3 equalling 18,7 øre/kWh)
Electricity prices (Danish Energy Agency, paper of April 1986)
Effective sales price incl. power contribution etc.: 31,4 øre/kWh.

Price of heat.

The price of heat is fixed based on the alternative energy cost of generating heat with gas fired boilers.
The price of gas being about 26 øre/kWh and the estimated annual boiler efficiency about 90% the alternative cost of heat is 29 øre/kWh.
To this price of energy a contribution is added which corresponds to investment and servicing costs otherwise connected with alternative heat production.
Investment in a new gas fired boiler plant is estimated at 1987 prices to amount to about 6,1 mill. DKK.
Savings in maintenance of existing heating centrals are estimated to amount to about 170.000 DKK in the fiscal year 84/85.
Operation costs are expected to be similar to the present costs at the existing district heating plant and are estimated to be Dkr. 410.000.
The consumption of electricity at the CHP-plant for pumps and ventilators will be app. 150 MWh/year. The consumption for the heat pump has been reducted from the electricity production.
The annual maximum of 7500 operating hours has been assumed. 660 hours are expected to be necessary for the planned maintance, preferably in the summer. The engine is assumed operated at 10% load in maximum 1000 hours/year and max. 90% in the remaining period.

Maintenance costs.

The maintenance costs are based on BWSCs experience of similar plant types (diesel engines) and international statistics.

Estimated costs of traditionel solution.

The CHP-plant in Hundested is to be compared with a conventional solution of a separate boiler plant (3 x 4 MW) and part of a centralized power plant (2.34 MW).

Based on former investigations in Hundested a new gas fired boiler plant with 3 x 4 MW boilers would cost approx. 6.1 mio DKK.

A 350 MW coal fired condensing type electricity plant would cost approx. 5000 DKK/kW.

The avoided cost in the public electricity system due to a CHP-plant of 2.34 MW electricity amounts to approx 11.7 mio DKK.

The total costs of a traditional solution would thus be approx. 17,8 mio DKK.

The extra cost involved in a CHP-solution is on the basis of above approx. 13.6 mio DKK excluding test and measurement costs of 0.2 mio DKK.

REFERENCE:

1 Loeth Per, Sectional Engineer,
Electricity Company EFFO Undalsvej 3
DK 3300 Frederiksværk, tel. +++45 42 12 02 10

TECHNICAL DATA.
RATED POWER

The rated power of the CHP plant is expected to be 2.34 MW-e and 3.92 MW-heat.

Annual output/consumption.
The expected annual production and consumption figures are listed below for average season (1993):

Expected production/consumption per year.

CHP-unit:	Prod. Electricity	: 13.706.000 kWh		
	Prod. Heat	: 22.960.000 kWh		
	Cons. Gas	: 34.929.000 kWh	3003	toe
	Cons. oil	: 3.037.000 kWh	261	toe
Boost fired boiler:	Prod. heat:	5.141.000 kWh		
	Cons. gas :	5.246.000 kWh	451	toe
Stand-by boiler:	Prod.heat :	2.749.000 kWh		
	Cons. gas :	3.054.000 kWh	262	toe

Substituted energy consumption.

Coal at condensing power station: 34.265.000 kWh 2.946 toe
Gas at conventional heat plant: 34.278.000 kWh 2.947 toe

Saved eq. oil consumption: 1.916 toe.

CHP-plant, Hundested: Annual operation budget.

```
CHP unit. nom performance:  3,92 MW-heat (60,4% of energy supplied)
                            2,34 MW-elec (36,1% of energy supplied)

Annual production/consumption:
       CHP-unit:
       Annual heat production          :22.960 MWH
       Annual elec production          :13.706 MWH
       Annual energy consumption       :37.966 MWH(8%oil-62,6% taxed)

Boost fired boiler:
       Annual heat production          : 5.141 MWH (98% efficiency.)
       Annual gas consumption          : 5.246 MWH (100% taxed)

Stand-by boiler:
       Annual heat production          : 2.749 MWH (90% efficiency.)
       Annual gas consumption          : 3.054 MWH (100% taxed)
```

<u>Annual income:</u> in 1000 DKK

```
       Electricity sold:    13.706 MWH a 314 DKK   : 4.304
       Heat sold:           30.850 MWH a 290 DKK   : 8.947
       Heat consumer contributions:
          Operation                                :   457
          Maintenance                              :   170
       ____________________________________________________

       Annual income                               :13.878
       ____________________________________________________
```

Annual expenses:

Item	Quantity and price		Amount
Energy costs:			
Gas (taxed)	:	30.165 MHW a 257 DKK	: 7.752
Gas (not taxed)	:	13.063 MWH a 74,5 DKK	: 973
Total: 3,93 mill. m3			
Oil (taxed)	:	1.901 MWH a 317 DKK	: 603
Oil (not taxed)	:	1.136 MWH a 130 DKK	: 148
Total: 307 m3			
Annual energy costs			: 9.476
Operation and maintenance costs:			
Operation staff (2 operators)			: 410
Maintenance (spare parts + assistance)			: 673
Lubrication oil (app. 14.800 litres)			: 148
Aux. elec. sonsumption: 150 MWH a 314 DKK			: 47
Operation and maintenance costs			: 1.278
Annual revenue (excluding capital costs)			: 3.124

Fig. 1

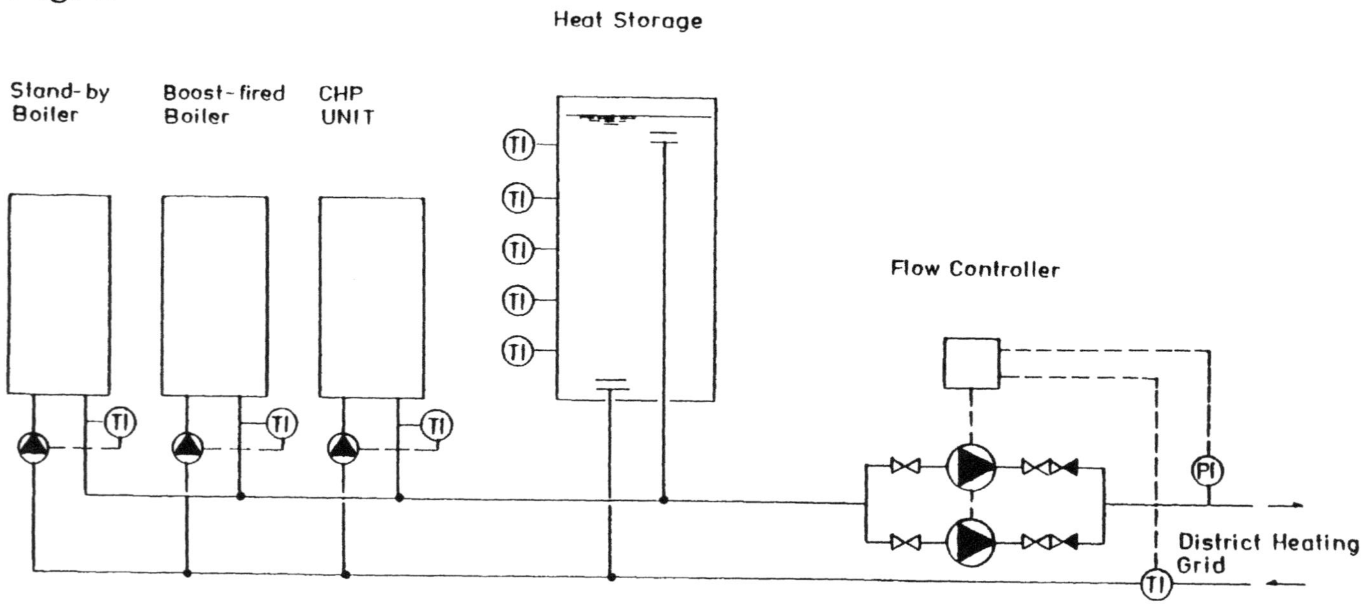

HUNDESTED:
Dristrict Heating System

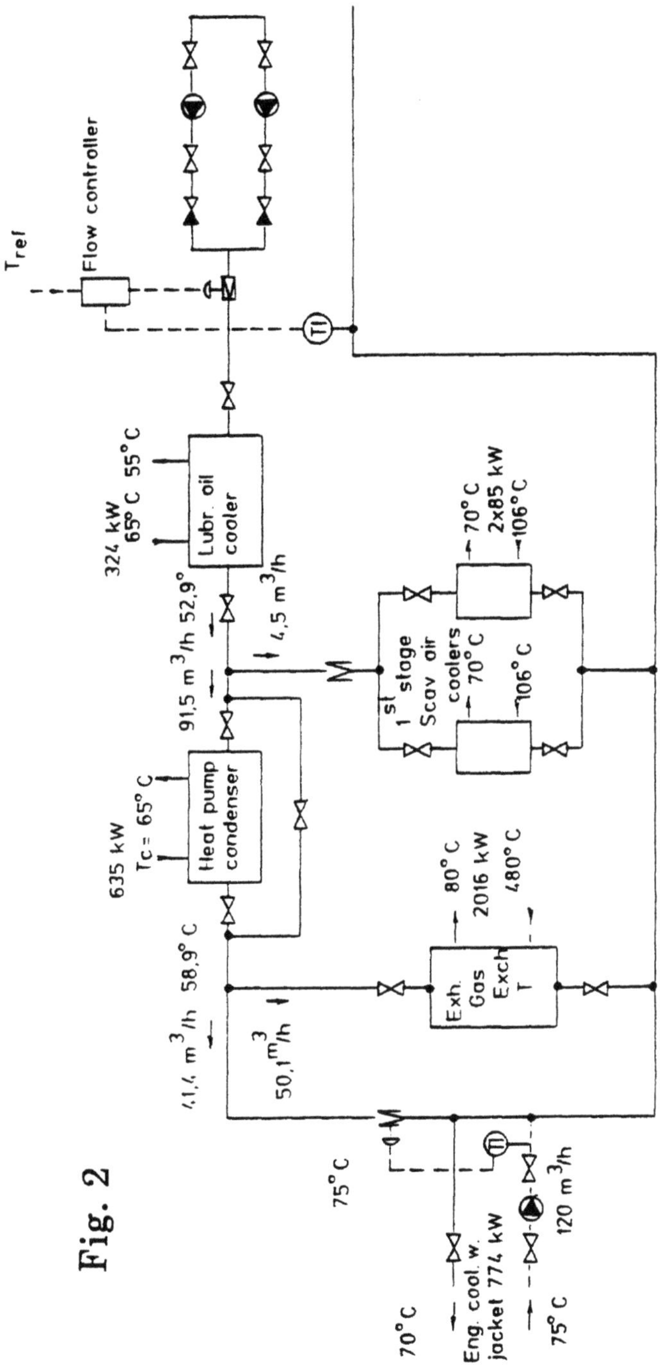

Fig. 2

HUNDESTED
High Temperature Cooling

Fig. 3

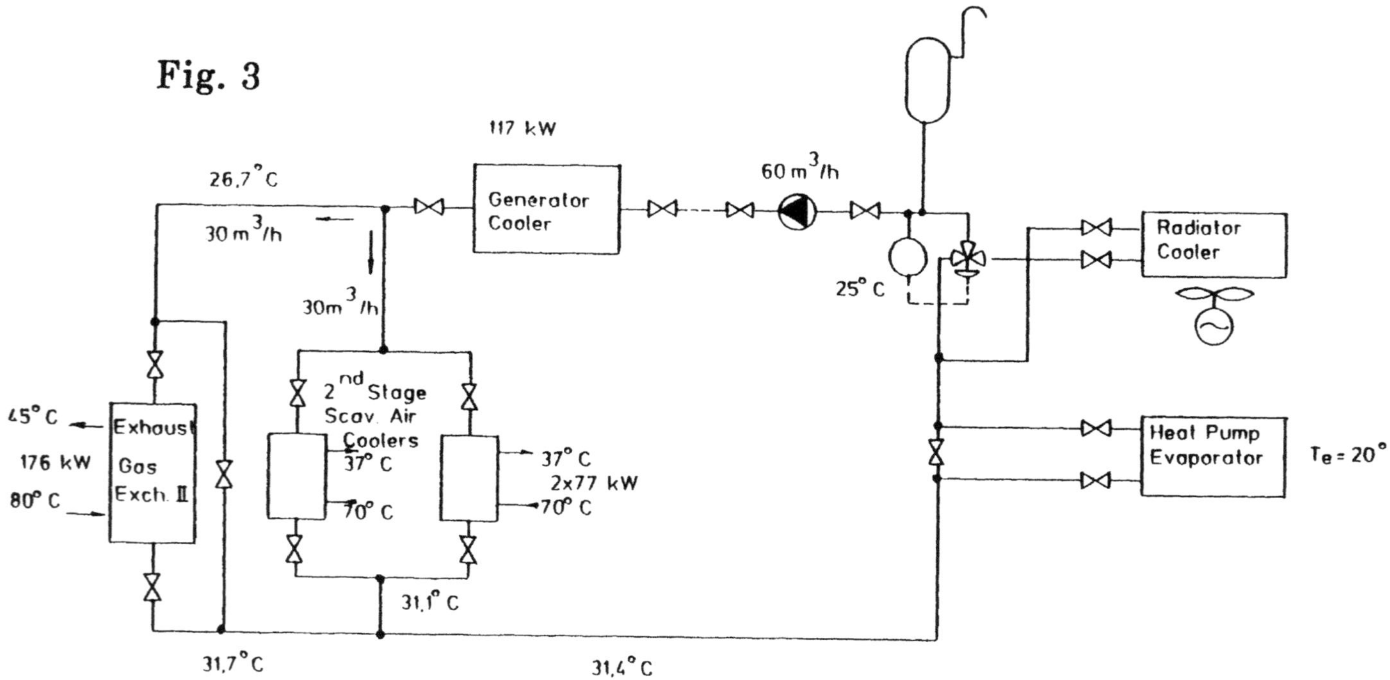

HUNDESTED:
Low Temperature Cooling System.

Fig. 4

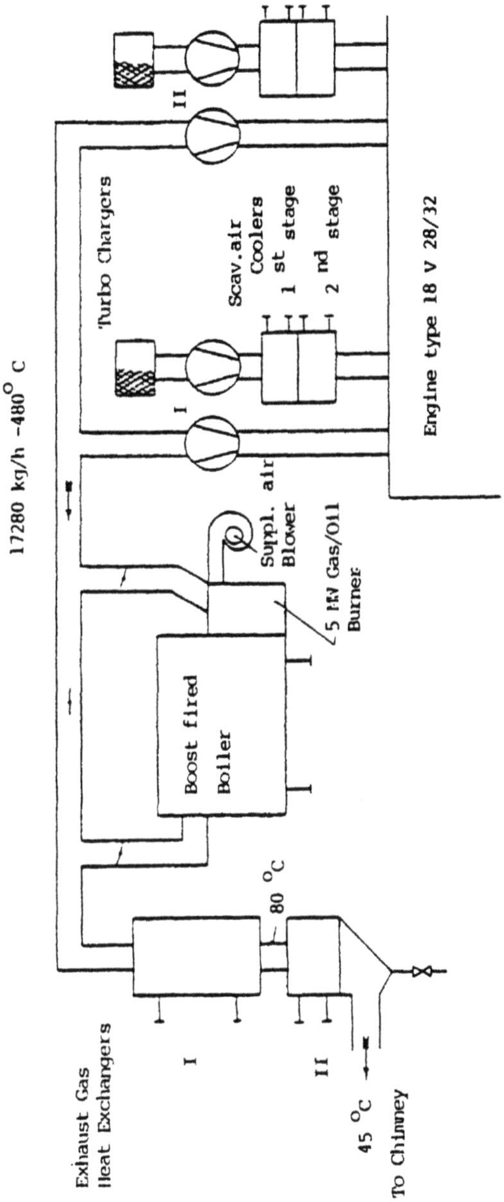
17280 kg/h -480° C
Turbo Chargers
II
I
Scav.air
Coolers
1 st stage
2 nd stage
Engine type 18 V 28/32
Suppl. air
Blower
5 MW Gas/Oil
Burner
Boost fired
Boiler
80 °C
Exhaust Gas
Heat Exchangers
I
II
45 °C
To Chimney

Fig. 5

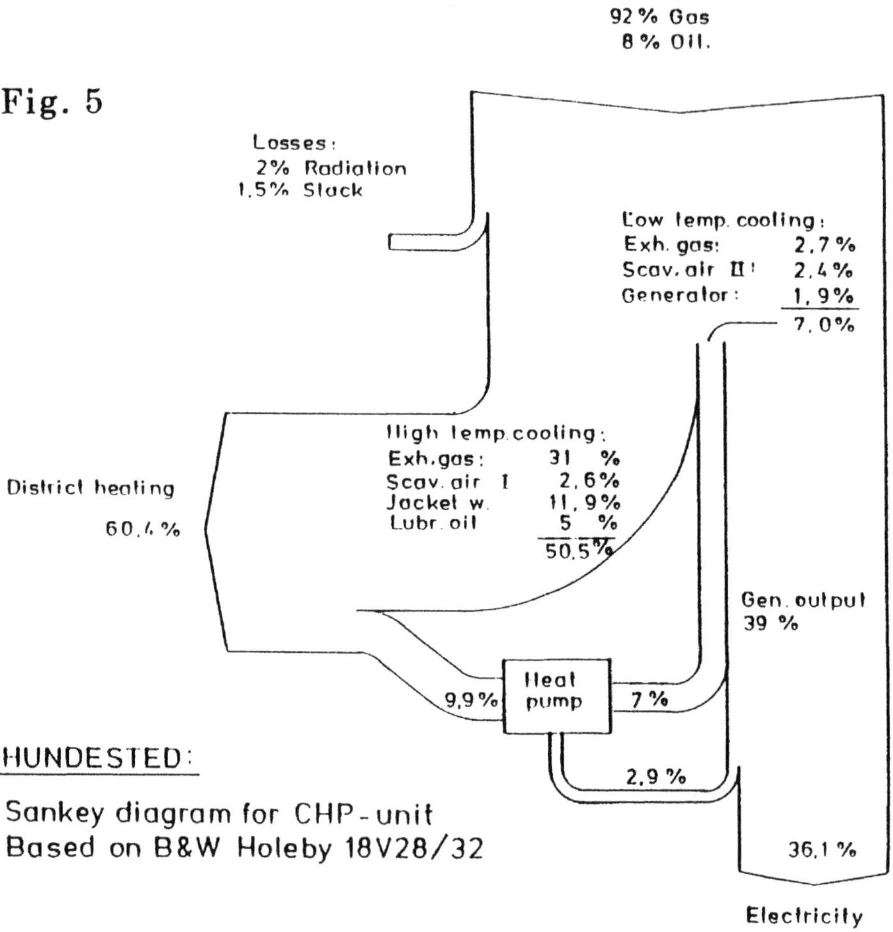

HUNDESTED:

Sankey diagram for CHP-unit
Based on B&W Holeby 18V28/32

CENTRAL 9.34 MW ELECTRICITY, HEATING AND COOLING COGENERATION PLANT

CARLOS FOUNAND COLL
LUIS MONTALT ROS

Nurel S.A. Spain

SUMMARY

This paper describes the optimum planning for the installation of a 9.34 MW gas turbine to generate electric power by recovering the heat generated by exhaust gases in the form of thermal energy -steam and hot water- and transforming part of this energy for the production of chilled water required by the plant's air conditioning system.

This project, which integrates diff.erent types of energy, has been declared by the European Economic Community as a Demostration Project within the area of energy savings and has been given a 900,000 ECU grant. The Ministry of Industry has also granted aid. This type of plant can operate continuously year-around.

RESUMEN

En esta ponencia se describe el óptimo de la instalación de una turbina de gas para generación de energía eléctrica de 9,34 Mw de potencia recuperando en forma de energía térmica -vapor y agua caliente- el calor de los gases de escape y utilizando parte de esta energía térmica en la producción de agua fría necesaria para el sistema de aire acondicionado de la planta.

Este proyecto, que integra diferentes tipos de energía, ha sido declarado por la Comisión de las Comunidades Europeas "Proyecto de Demostración" dentro del campo del Ahorro Energético, habiéndose concedido una subvención de 900.000 ECUS. También ha recibido ayudas del Ministerio de Industria. Este tipo de plantas puede funcionar de forma contínua todo el año.

CENTRAL 9.34 MW ELECTRICITY, HEATING AND COOLING COGENERATION PLANT

Carlos Founaud Coll

Doctor in Industrial Engineering - Head of Engineering, NUREL, S.A.

Luis Montalt Ros

Doctor in Industrial Engineering - Project Manager, NUREL, S.A.

PROJECT DESCRIPTION

NUREL, S.A. is an industrial plant belonging to the ICI Group and is engaged in the manufacture of synthetic fibres, nylon and polyester. It is located in Zaragoza, km 329, Carretera Barcelona.

The plant is fitted with polymerisation and spinning facilities for both fibre types and as these are continuous, highly reliable ancillary services are required, such as the supply of steam, electricity and water, air conditioning and nitrogen.

Energy conservation and management policies have led to the study and completion of a cogeneration plant with the following basic characteristics:

- Natural gas turbine with a 9.34 MW generator, equipped with air evaporative cooler and gas compressor.
- Recovery boiler using the heat of the gases discharged from the turbine at a temperature of 450-500°C to produce 21 T/h of steam and afterburner allowing peak rates of up to 30 T/h to be served.
- Recovery of flue-gas residual heat to obtain hot water, in addition, representing a power source of 2100 kJ/s with flue-gas discharge into the atmosphere at 90-95°C.
- Absorption cooling plant using lithium bromide as refrigerant and consisting of:
  - One 400 T cooling unit, using recovered hot water as heat source.
  - Two units, 750 TR each, using low pressure steam as heat source.
  - All the above units produce chilled water for the air conditioning system with an 11.5 head at 4.5°C.

- Renewal of the entire 6.3 kV installation and equipment feeding the plant. This was necessary as a result of the higher short-circuit power involved since the generator will operate in synchronisation with the distribution system of the utility Electricas Reunidas de Zaragoza (ERZ, S.A.) It has been estimated that the electricity production will equal 75.80% of the plant's needs.

- Distributed control system with redundant configuration to improve reliability.

ENERGY CONSUMPTION

The attached graph shows the steam use rates and gives the monthly process steam supply for 1989 and the forecast for 1990 as a result of internal changes.

With this particular energy use level, a 6 to 7 MW generating plant could be justified. However, if the new steam requirements for the absorption plant are added, it is justified to step up the plant power to 9.34 MW as designed.

A reduction in consumption can be seen in the electricity consumption rates graph, which is more marked in summer time as a result of the chilled water compressors being stopped and replaced by the absorption plant.

ENERGY CONSERVATION

Steam

Steam consumption for 1989 and the following years, taking into account the planned expansion of the cut fibre process, will give a gas requirement of the existing boilers of 97,037,000 Therm/year, equivalent to 406 Tera-Joules/year (PCI).

Gas Turbine with Synchronised Generator

Operating year-round continuously at nominal load values, les one down period for maintenance, the result would be:

- Natural gas consumption................ 918 TJ/yr
- Electric power production............. 76.8 GWh/yr

Thermal Energy Use by Absorption Units

Taking into account the operating time of these units to cover the demand of chilled water, particularly during the hot season, by replacing the current freon gas compressors, the units will require thermal energy in the form of steam and hot water in the order of 120 TJ/yr.

Electric Power Demand

In the present conditions, the plant draws 94.6 GWh/yr. The future situation with cogeneration will be:

- Present consumption....................94.6 GWh/yr
- Less consumption chilling compressors. 5.7 GWh/yr
- Plus auxiliary cogeneration units..... 0.7 GWh/yr

- Total demand..........................89.6 GWh/yr
- Cogeneration production...............76.8 GWh/yr

Total demand from ERZ, S.A. ....12.8 GWh/yr

Gas for Combustion

Considering the existing boilers will be used while the cogeneration unit goes into yearly maintenance downtime.......15 TJ/yr

Also the afterburner for the recovery boiler, when the predicted steam needs so require, particularly in summertime...... 9.4 TJ/yr

TOTAL.....24.4 TJ/yr

Energy Saving

	Current	With Cogeneration
Natural gas consumption TJ/yr........	406	942.4
Electricity generated by the conventional gas-burning plant (PCI).......	981	132.7
Total............................	1,387	1,075.1
Energy savings.......................	312 TJ/yr	(7,610 Tep)

This would be the energy savings for Spain originated by the cogeneration project being presented here.

A conventional power plant performance has been considered at 2,500 kcal/kWh, in PCS, equivalent to using natural gas as fuel according to the Order of 7 July, 1982 of the Ministry of Industry (BOE, 17 July, 1982). However, if any other type of fuel is considered, such as coal, lignite, fuel oil, etc., the above figure would be considerably higher and the savings proportionally greater.

INVESTMENT

The investment costs originally estimated were 1,126 million pesetas.

ENERGY CONSUMPTION AND COST BY UNIT OF PRODUCTION

The yearly plant production, including all types of fibre, is 29,647 t/yr.

To obtain this production the following amounts of energy are used (without cogeneration):

- Electricity......................94,6 GWh
- Natural gas as fuel.............97 x 10^6T

With the cogeneration system, the energy consumption will be:

- Electricity from grid...................12.8GWh
- Natural gas for turbine and fuel 225 x 10^6 Therms

Every consumption by kilo of product will then be:

	kWh/kg	GM Therm/kg
.Without cogeneration..........	3.191	3.27
.With cogeneration.............	0.43	7.60

Applying energy costs per kilo, both current and future with cogeneration, we have:

. Energy costs without cogeneration......30.80 Pts/kg.

. Energy costs with cogeneration.........18.10 Pts/kg.

. Savings due to cogeneration............12.70 Pts/kg.

These savings cannot be considered as net savings since the new facilities involve increased maintenance costs above present costs estimated at 0.60 Pts/kg and thus the final savings will be in the order of 12.10 pts.Kg.

After a three to four-year period has elapsed, savings will be reduced to 9.80 Pts/kg. due to changes in energy prices.

The investment payback period can be estimated at 3.2 years.

This cost reduction will allow Spanish fibres to be more competitive in the European market.

PRESENT PROJECT STAGE

The project is currently at a very advanced stage of completion. Although it was originally planned to begin continuous energy production on 1 January, 1990, delays in delivery of important equipment will force the commissioning date to be postponed to 1 March, 1990.

The project cost will go beyond the original estimations and will come close to 1,200 million pesetas.

ACKNOWLEDGEMENTS

Nurel, S.A. wishes to express its gratitude for the support and co-operation given the execution of this project to the:

- EUROPEAN ECONOMIC COMMUNITY
  (DIRECTORATE GENERAL OF ENERGY)

- I.D.A.E.
  (INSTITUTE FOR ENERGY DIVERSIFICATION AND SAVINGS)

- THE SPANISH MINISTRY OF INDUSTRY
  (DIRECTORATE GENERAL OF ENERGY)

- DIPUTACION GENERAL DE ARAGON
  (REGIONAL COVERNMENT)

- ELECTRICAS REUNIDAS DE ZARAGOZA, S.A.

- ENAGAS, S.A.

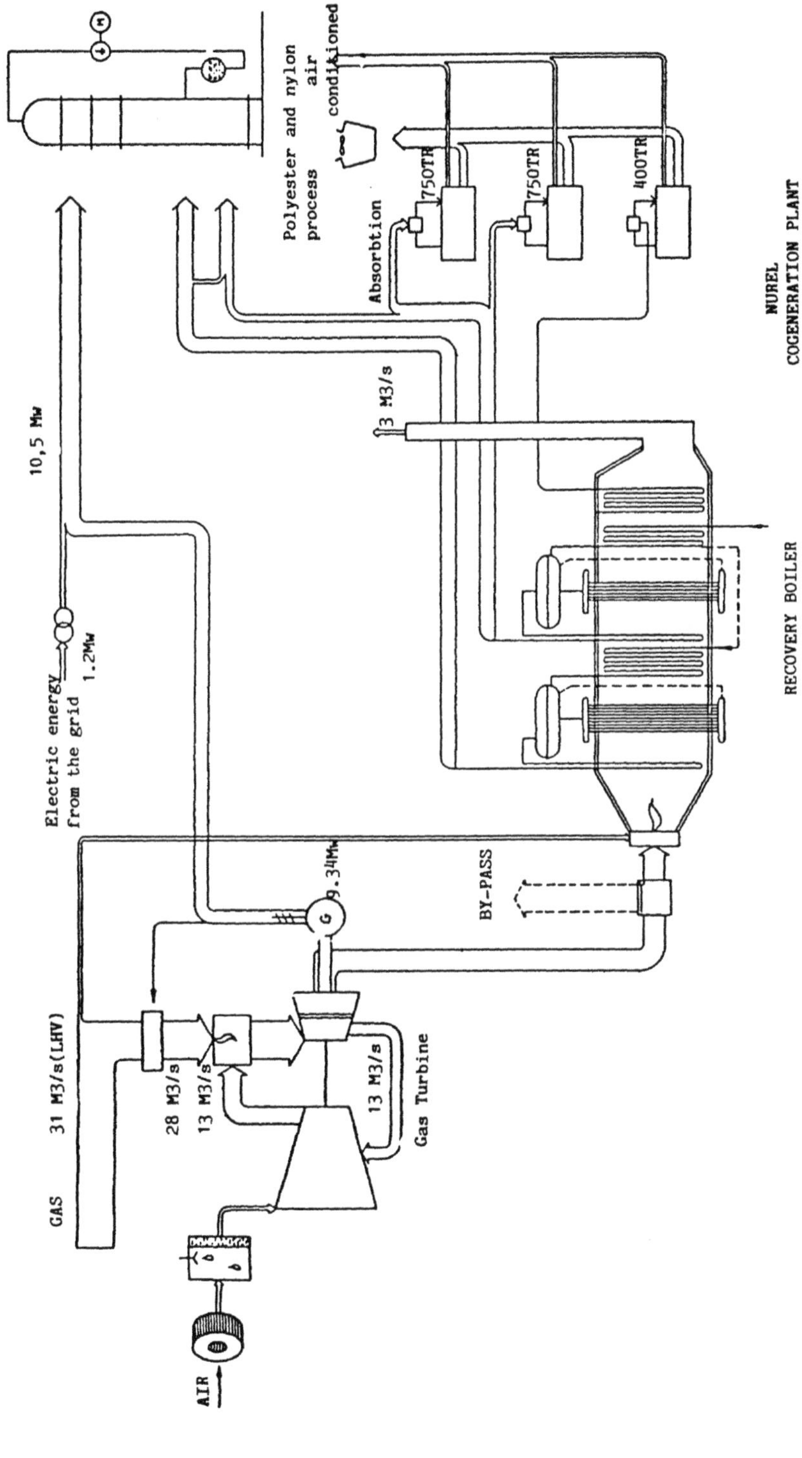

NUREL COGENERATION PLANT

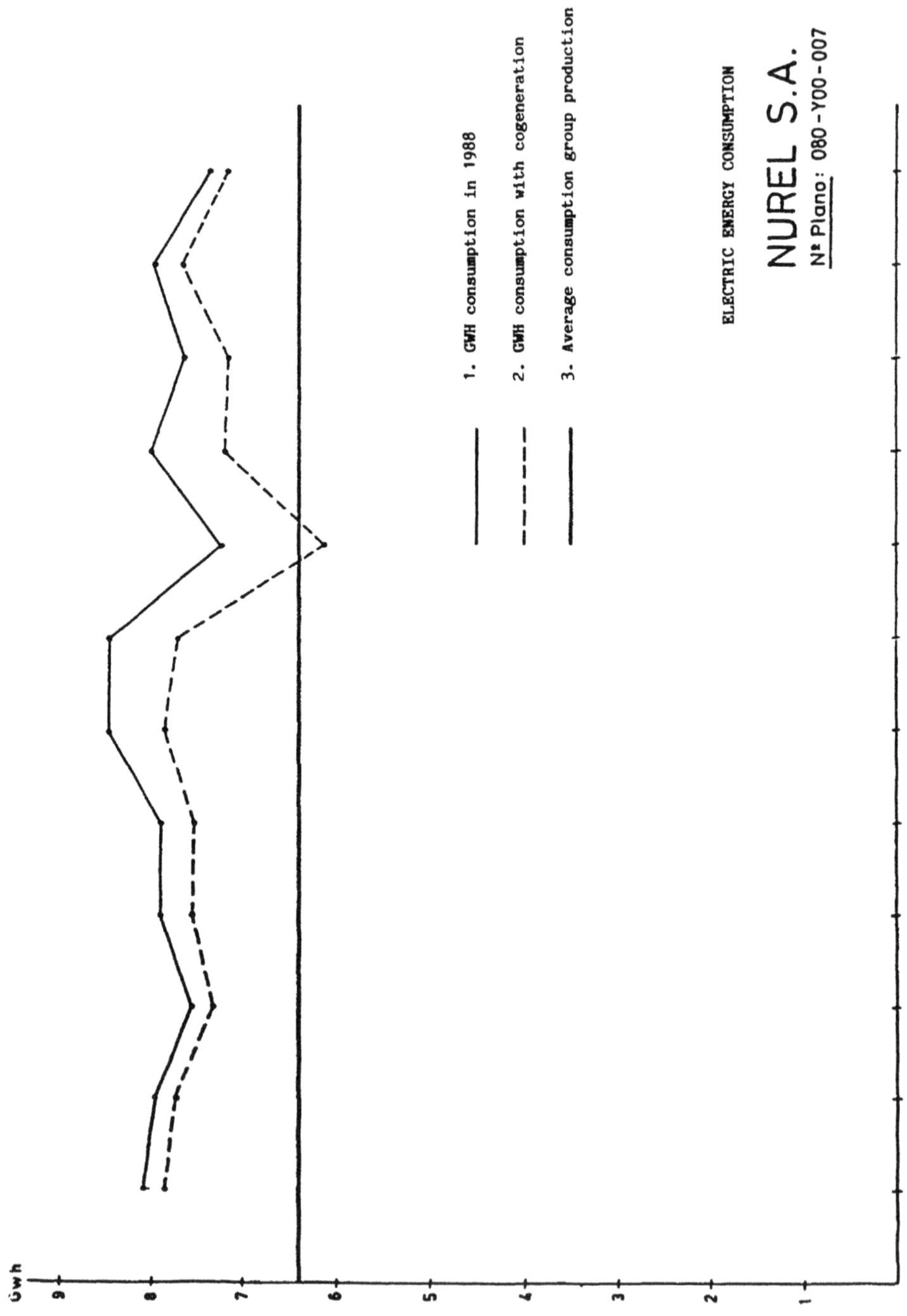
Gwh
9
8
7
6
5
4
3
2
1
ENERO
FEBRERO
MARZO
ABRIL
MAYO
JUNIO
JULIO
AGOSTO
SEPT.
OCT.
NOV.
DIC.
AÑO 1988
1. GWH consumption in 1988
2. GWH consumption with cogeneration
3. Average consumption group production
ELECTRIC ENERGY CONSUMPTION
NUREL S.A.
Nº Plano: 080-Y00-007

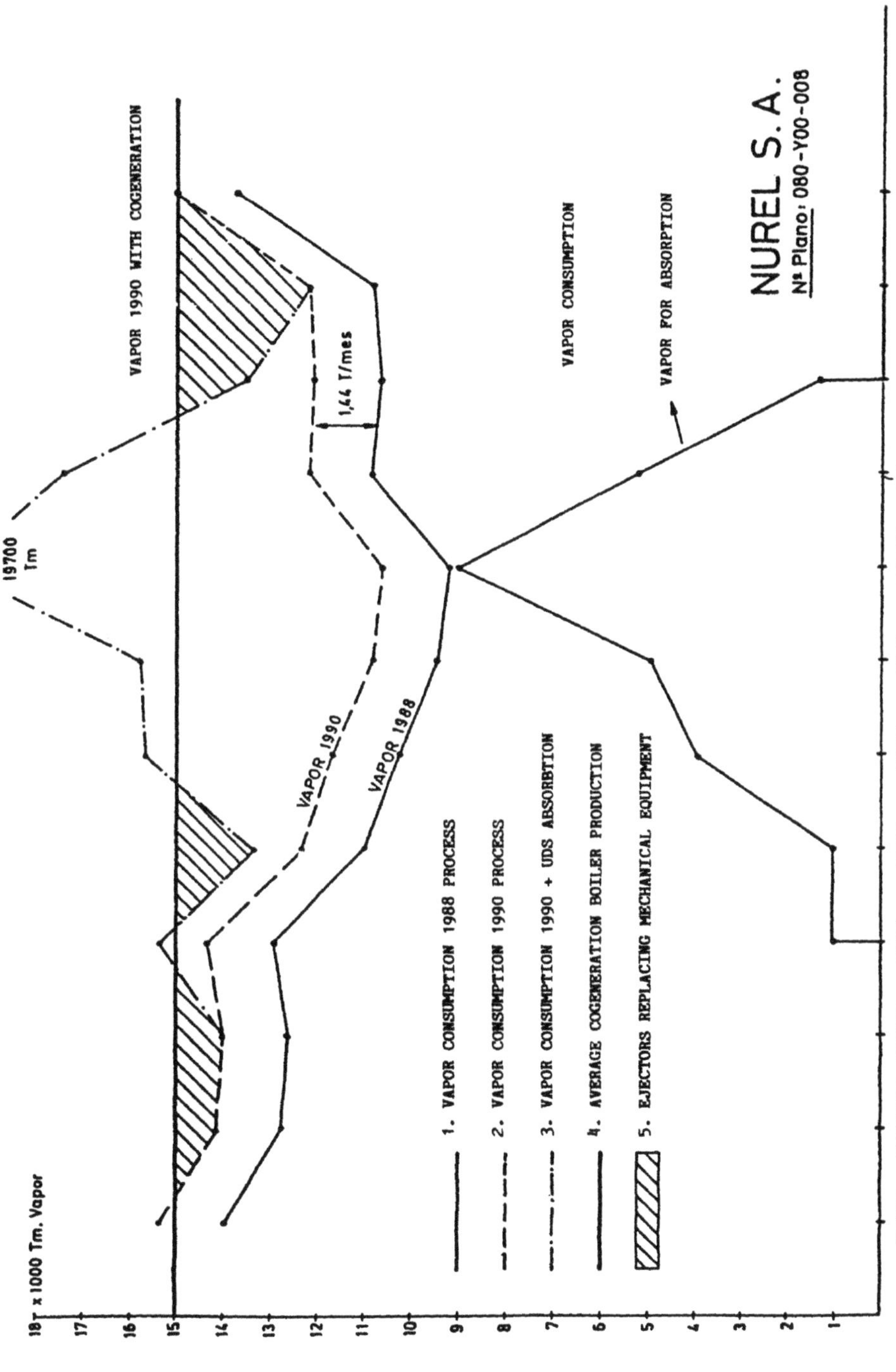
x 1000 Tm. Vapor
VAPOR 1990 WITH COGENERATION
19700 Tm
1,44 T/mes
VAPOR 1990
VAPOR 1988
VAPOR CONSUMPTION
VAPOR FOR ABSORPTION
NUREL S. A.
Nº Plano: 080-Y00-008
1. VAPOR CONSUMPTION 1988 PROCESS
2. VAPOR CONSUMPTION 1990 PROCESS
3. VAPOR CONSUMPTION 1990 + UDS ABSORBTION
4. AVERAGE COGENERATION BOILER PRODUCTION
5. EJECTORS REPLACING MECHANICAL EQUIPMENT
ENERO FEBRERO MARZO ABRIL MAYO JUNIO JULIO AGOSTO SEPT. OCT. NOV. DICI.
AÑO 1988

COGENERATION IN EUROPEAN COMMUNITIES' MEMBER STATES

DISCUSSION

SUMMARY

PARTICIPANTS

The following participants have asked questions or made commenrts:

GRANER, Solvay (Spain); ECHEVARRIA, Michelin (Spain); JACUBOWIEZ, I., Elf Aquinaine (France); SANCHEZ, Sevillana de Electricidad (Spain); SOLIS, A., Mecánica de la Peña (Spain); RUIZ VALDEPENAS, A., Cristalería Española S.A. (Spain); FERNANDEZ ZORRILLA,A., Iberduero (Spain).

SPEAKERS

CAPARROS, J.J., Papelera del Jarama (Spain); MARANIELLO,G., Ansaldo (Italy); BERKELMANS, F.W., Royal Schelde (The Netherlands); Mc GLADE, P.P., Pfizer Chemical Corp. (Ireland); HOLLTROTTER, E, DOW Stade GmbH (Germany); LOETH, P., Elselskabet EFFD (Denmark) and MONTALT ROS, L., Nurel S.A. (Spain).

TOPICS DISCUSSED

- The flue gas cleaning and afterburners at the Duiven plant.
- The flue gas temperature at the entrance of the second pass and composition of the flue gases from the furnace in the Dutch plant.
- H_2/ CH_4+ H_2 ratios in gas turbine recycling.
- Investment and operating costs of the Nurel plant.
- Turbogenerator investment figures for the Nurel plant.
- Grid storm disruptions and cogeneration liability.
- Single and double effect absortion machines.
- Total investment in cogeneration plants.
- Forecasted and real load factor in the Papelera de Jarama's plant.
- Use of the refrigerator's water heat in the Ansaldo plant.

COMMENT

The discussion was focused on very specific technical details of the session's projects. Most, if not all of the questions and comments are actually widely covered in the speaker's papers. It may be worth noting that, for example in the case of Papelera del Jarama, the reliability of the cogeneration system is highly superior to the one of the grid. In fact, one of the reasons to have converted to cogeneration was the need for higher reliability in storm grid distruption situations.

CONCLUSIONS

Mister Chairman, Ladies and Gentlemen,

let me now summerize and draw the conclusions of this seminar:

During the **opening session**, Mr PEREZ PRIM, Director General for Energy at the Spanish Ministry of Industry and Energy, has talked on the commitment of the Spanish Government - along the lines of the European Commission - with cogeneration. He has announced that national decisions on a new legislation could be taken in the very short term.

Mr KINDERMANN, Head of Division at the E.C. Directorate General for Energy, has made a general overview of the Commission's programme in the field of energy conservation in the framework of the Common European Energy Policy.

Mr SERRANO, Director General of I.D.A.E., made an overview of the cogeneration in Spain and the I.D.A.E. involvment in the field.

During the **second session** on overviews of technologies, Mr ALBISU (SENER, Spain) gave us an overall review of the principles on which the interest in and possibilities of cogeneration are based. He compared different alternatives and their main results.

Mr GYFTOPOULOS (M.I.T., USA) made a wide survey of already made small and medium size projects of cogeneration with special examples of wood and biomass fueled systems.

Mr CONTRERAS and Mrs GOLEZ ANGULO (I.D.A.E., Spain) made a detailed analysis of the technology possibilities, the legal framework and the financial solutions of cogeneration projects in Spain, in the past, presently and in the future.

During the **third session** on financing and legislations, Mrs HAMRIN (Independant Energy Producers Ass., USA) made an in depth presentation of the conditions necessary for the development of cogeneration: financing, projects, risks, general constraints, environmental implications, with special references to the State of California.

Mr FEE (European Commission) has spoken on the rôle of Commission and member states in promoting Energy Saving Companies (ESCOS) and in novel financing mechanisms. He has described the work of the Commission in this field.

Mr DRISCOL (I.E.A.) has given an overview of the legal obstacles to cogeneration in ten non E.C. industrial countries.

The **fourth session** was a highly motivating round table with many interesting questions from the audience on "cogeneration and environment".

Have taken part in the round table:

Mrs HAMRIN, MM. DIAZ VARGAS, DRISCOL, FEE, GREEN, GYFTOPOULOS and SIRCHIS.

If we would like to summarize this round table, I think that a main conclusion can be drawn: Cogeneration is the way in these changing times because it increases the energy efficiency and, at the same time, it helps to improve the overall environment.

During the **fifth and last session** on "Cogeneration demonstration projects", we have heard of plants in paper and chemical industries, district heating and theoretical and practical technology achievements. It is not possible to summarize the seven communications delivered this morning on these very interesting projects in several sentences. I am convinced that we all have learned much and that the technical and financial solutions proposed will help our countries to be more efficient in energy, more environmental safe and to increase the competitivity and the profitability of our enterprises.

Mister Chairman, Ladies and Gentlemen, it remains now to me, before the closing of the seminar to thank you all on behalf of the organizers for having attended the Seminar and for your attention. I thank all the speakers and those who have taken part in the discussion.

May I finally on behalf of you all thank our interpreters and the local organizers of I.D.A.E., more specially Mrs A. GARCIA, Mrs A. GONZALES MONFORT and Mr ARIMANY.

J. SIRCHIS

LIST OF PARTICIPANTS

A.Mitja I.S.
G.De Cataluna
Avd.Diagonal 514 2N
E - Barcelona 08006

A.Nieto
Sercobe
Espana

A.Sepulveda R.
Rosesi s.a.
Profesor Waksman 12
E - Madrid 28036

Alberto Sanchez
Sereland
Aribau 200-120
E - Barcelona 08036

Alvaro Montelay
La Salvadora s.a.
Mayor 60
Guipuzcoa
E - Villabona

Alvaro Villar
Idae
Paseo de la Castellana 95 planta 21
E - Madrid 28046

Amadeo Martorell
Sereland s.a.
Aribau 200
E - Barcelona 08036

Angel Cnacer
Renfe
Avda Pio XII S/N Caracolas 10
E - Madrid

Angel Mujica O.
La Salvadora s.a.
Mayor 60
Guipuzcoa
E - Villabona

Angel Sancho Ros
GHSA
Concha Espina 63 5 planta
E - Madrid

Antonio Olbes R
Repsol Petroleo
Jose Abascal 4
E - Madrid 28003

Antonio Suarez F;
Servicios Energeticos s.a.
Balmes 262 6 planta
E - 08006 Barcelona

Arnovil Guy
Compagnie Générale de Chauffe
63 rue de Gerland
F - 69363 Lyon Cedex 7

Artal C.
Empresa Natonal de Celulosas
Juan Bravo 49 DPDO
E - 28006 Madrid

Beauduin G.
Shell Recherche s.a.
F - 76530 Etrand Couronne

Bent Johannesen
Trekantomradets
Varmetransmissionsselkab 1/S
Tonne Kjersvej 11
DK- 7000 Frederica

Berkelmans F.W.
Royal Schelde
Holland

Bernardo Caso
Idae
Paseo de la Castellana 95 planta 21
E - Madrid 28046

Besch H.
Fernwarme Verbund Saar Gmbh
Bismarckstrasse 11
D - 6620 Volkingen

Bourgedis B;
Total Compagnie Franc. des Pétrol
24 rue Erlanger
F - 75016 Paris

Buuren J.E.

Warande 17
The Netherlands

C.Foundaud Coll
Nurel
Barrio Santa Isabel Malpica
E - Zaragoza 50016

C.Ibar Klingber
Direccion Genral de la Energia
S - Estocolmo 11787

C.Moreno Serrano
Union Electrica Fenosa
Capitan Haya 51
E - Madrid

Carlos Perea E
Iberduero
Gardoqui 8
E - Bilbao 48008

Carlos Perea E.
Iberduero
Gardoqui 8
E - Bilbao 48008

D.Sanchez Mena
Sevillana de Electricidad
Espana

D.Van Der G.
ESTS bv
Adrescode 2H 14
NL - Velsen Noord 1951 JZ

David Jhons
Hawker Siddeley Power Ing.Ltd
Santa Engracia 3 4izq
E - Madrid

Dimitri V. Papaconstantinou
Public Power Co
32 Arachovis str.
GR - 10681 Athens

Dr Alfred Reichl
Verband der Elek.Osterreichs
Brahmplatz 3 Postfach 3
A - 1040 Wien

Dr Carlos L.Lopez Cacicedo
Elect.Council Research Centre
UK - Capenhurst Chester CH1 6es

Dr Eulogio Reinoso
Empesa Asociacion de Energia Electr
Hermosilla 31
E - 28001 Madrid

Dr Georg.Alefeld
University of Munich
19 James Franck Strasse
D - 8046 Garching B. Munchen

Dr J.Muller
Jura Cement Fabriken
CH - 5103 Wildegg

Dr Kurt Fleckenstein

Adernauerallee 148
D - 53000 Bonn 1

Dr P.Diervem
S.C.K.
Boerentang 200
B - 2400 Mol

Dr Reinhard
Planung Energie und Vautechnik Gmbh
Passauer strasse 8/9
D - Berlin 100

Dr Reinhard J.
Energie und Bautechnik gmbh
D - 1000 Berlin 30

Dr Valkanas
University of Athens
42 Patision str.
GR - 10682 Athens

Dr. Ing. Fritz Pfisterer
University of Stuttgart
Pfaffenwaldring 47
D - 7000 Stuttgart 80

Dr. Turberfield KC
Harwell Laboratory B 329
UK - Oxfordshire OX11 ORA

Dubbelo M.

P.B. 13766
NL - 2501 ET Den Haag

Duran P.
Instituto de Ceramica y Videro
F - 64500 St Jean de Luse

E.Carballo G.
Forjas y Aceros de Reinosa s.a.
Paseo Alejandro Calonge 1
E - Cantabria Reinosa

E.Iglesias
Union Electrica Fenosa
Capitan Haya 51
E - Madrid

E.Puig Lopez
Telsa
plaza Eguilaz 7
E - Barcelona 08017

Emilio Rubio A.
Constructora Equipos Electricos
Apartado 1096
E - Bilbao 48080

Escondeur M;
Const.Dirigeables Pays Basque
9 allée des Fleurs
F - St Jean de Luz

Esteban Diez
Idae
Paseo de la Castellana 95 planta 21
E - Madrid 28046

Evans E.C.
Confederation of British Industry
103 New Oxford str
UK - London WC1 A1DU

F.Albisu
Sener s.a.
Espana

F.Diaz Caneja
Unesa
Espana

F.Domingo M.
Enagas
Avda America 38
E - Madrid

F.Torija Mtnez
Balcke Durr Espanola s.a.
Orense 81 1'
E - Madrid 28020

F.Zap

Samaria 14
E - Madrid 280

Felix Gutierrez
Indeim
Guzman en Bueno 133
E - Madrid 28003

Fernando Alegria
Esc.Tec.Sup.Ing.de Minas
Rios Rosas 21
E - Madrid 28003

Fernando Belaso
Autonomo
Corregidor Juan Fransisco L. 34
E - Madrid 28030

Fernando Lopez P
La Seda de Barcelona
Ctra de Camarma KM 2,8
E - Alcala de Henares Madrid

Fournier

47 av. Laplace
F - 94117 Arweil cedex

Fransisco Mateu
Servicios Energeticos s.a.
Balmes 262 6 planta
E - Barcelona 08006

Fransisco Saenz
Trade & Service International
Ganduxer 5
E - Barcelona 08021

Garcia Ana
Idae
Paseo de la Castellana planta 21
E - Madrid 28046

Geoffrey C. Angell
Holec Ltd
1/13 High str
UK - Leatherhead Surrey KT22 8AA

Georgikis Scholis
Philkeram Jomnson s.a.
P.O. Box 10213
GR - Thessaloniki 54110

Gimenez Tresaco
Hidroelectrica de Cataluna
Espana

Green D.
Combined Heat and Power Asociation
35/37 Grosvenor Edns
UK - London

Gregorio Sainz

Issac Peral 18
E - Madrid 28015

A.Gomez Angulo
Idae
Paseo de la Castellana 95 planta 21
E - Madrid 28046

Gunther Lubish
Dept.Environment and Energy
Styresemannstrasse 26
D - 4000 Dusseldorf 1

Gustavo Reimers
Indeven
Espana

Guyart
Technip Process
170 place Henri Regnault Cedex 23
F - 92090 Paris La Défense

H. Van Der B.
S.C.K.
Boeretanc 200
B - 2400 Mol

Hans Jorgen Koch
Ministry of Energy
Slotsholmsgade 1
DK - Copenhagen K 1216

Hans Olson
Direccion General de la Energia
S - Estocolmo 11788

Heikki Kauppi
Ivo International
Avda Republica Argentina 273-7
E - Barcelona 08023

Helmut Kron
Hoechst AG Energie
Postfach 80 03 20
D - 6230 Frankfurt a.m. 80

Hutin
Maitrise de l'Energie Cofreth
46 me Letort
F - 75018 Paris

J. Franco G.
Elecnor
plaza Ciudad de Salt, Bajo
E - Barcelona 28043

J.A.Gonzalez C.
ABB Energia
Ramirez de Arellano 17
E - Madrid 28043

J.A.Gonzalez R.
Initec
Alenza 4
E - Madrid

J.A.Guillon M.
Enagas
Avda America 38
E - Madrid 28028

J.Almela C.
Catalana de Gas s.a.
Corcega 373 5 planta
E - Barcelona 08037

J.Barrio S.
Forjas y Aceros de Reinosa s.a.
Paseo Alejandro Calonje 1
E - Cantabria Reinosa

J.Dominguez A.
Sevillana de Electricidad
Espana

J.E.Casado
Celulosas de Asturias s.a.
Apartado 39 Asturias
E - Navia 33170

J.E.Fuster C
Enagas
Avda America 38
E - Madrid 28028

J.Ferrer Mateos
Union Electrica Fenosa
Capitan Haya 51
E - Madrid

J.Garrido A.
Asoc. Nal. Ftes. Papel y Carton
Alcalada 85 4
E - Madrid 28009

J.I.Martinez Y.
Portland Valderrivas s.a.
Jose Abascal 59
E - Madrid 28003

J.L. Cabanas P.
Sociedad Nestlé A.E.P.A.
Cantabria
E - Penilla de Cayon 39650

J.M. Perez Prim
Ministerio de Ind.y Energia
Paseo Castellana 160
E - Madrid 28046

J.Manuel Lopez
s.a. Camp
Fray Carbo 24
E - 08400 Barc Granollers

J.Maria Manso
Consusa
Riere Las Paret
E - Barcelona 08850

J.Mauri Majos
S.Miguel Fab.Cerveza y Malta s.a.
Apartado 67
Lerida
E - Lleida 25080

J.Ramirez Card
Dragados y Construcciones
Espana

J.Roca Serradel
Catalana de Gas s.a.
Corcega 373 5 planta
E - Barcelona 08037

J.Solaun de G.
Derivados del Fluor s.a.
Onton Cantabria
E - Castro Urialdes

J.Terrades F.
IPEAE
Avellanas 14
E - Valencia 46003

J.Torras T.
Catalana de Gas s.a.
Corcega 373 5 planta
E - Barcelona 08037

Jaime Claramunt
Dir.Gnral.de Energia Generalitat
Diagonal 514 2 planta
E - Barcelona 08006

Javier Diaz
Power T.C.Espanola
Orense 5
E - Madrid

Javier Franco G.
Heredia y Moreno
Princesa 3
E - Madrid

Javier Fuentes
Hidroelectrica Espanola
Hermosilla 3
E - Madrid 28001

Jesus Cano M.
Repsol Petroleo
Jose Abascal 4
E - Madrid 28003

Joachim Vob
Metaleurop Wesesr Biel Gmbh
Johannastrasse 2/ Box 16
D - 2890 Nodenham

Jordi Farre
Gas Tarraconense
La Unio 21
E - Tarragona 43001

Jorge Baviera
Baviera Ahorro de Energia
Gomez Ferrer 33
E - 46900 Vale Torrente

Jorgen Overgaard
Danish Steel Works Ltd
DK - 3300 Frederiksverk

Jose I.Menendez
Taller de Ideas
Zurbano 74 1 Derecha
E - Madrid

Jose Luis Balboa
Cartos Espana s.a.
Ctra.Burgos Portugal KM 126
E - Valladolid 48008

Jose Luis Ortega
Enagas
Avd.America 38
E - Madrid 28028

Jose Maria Egea
Unigas
Paseo Castellana 123
E - Madrid

Josep Pueyo
s.a. Camp
Fray Carbo 24
E - Barcelona Granollers 08400

Juan Emanuel A
Esc.Sup.de Ing.Industriales
Alameda de Urquijo s/n
E - Bilbao

Juan Garay
Direccion Nacional Energia
S - Estocolmo 11787

Juan Manuel A
Inseriales
Artaza 7 Tercero
E - Vizc Leida 48940

Juan Martin Albo
Indein
Guzman el Bueno 133
E - Madrid 28003

Juan Ramon M.
Indein
Guzman el Bueno 133
E - Madrid 28003

Julian Deguez L
Ghesa
Concha Espina 63 5 planta
E - Madrid

Heliotronic gmbh
postfach 1129
D - 8263 Burghausen

Julian Mut Puch
Servicios Energeticos s.a.
Balmes 262 6 planta
E - Barcelona 08006

Julio Bayo R.
La Seda de Barcelona
Ctra.de Camarma KM 2,8
E - Alcala de Henares Madrid

Jurgen Schlag
Balcke Durr a.g.
Orense 81 1'
E - Madrid 28020

Karidogiannis E. P.
Public Power Co
32 Arachovis str.
GR - 10681 Athens

Katopodis G.
Asprofos Engineering s.a.
50 El.Venizelon ave.
GR - Athens Kallithe 17676

Klop P.G.
Regio Arnhem
Holland

L.Suarez B.
Enagas
Avda America 38
E - 28028 Madrid

Leandro Martinez
Elecnor
Plza Ciudad de Seita 4 Bajo
E - Madrid 28043

Leandro Martinez
Elecnor
Plza Ciudad de Seita 4 Bajo
E - Madrid 28043

Lefebrvre G.
Centre Tertiaire de l'Arsenal
299 rue St Sulpice B.P. 245
F - 59504 Douai

Lelorrain Ph.
Soc.Alsacienne dde Construct.Mecan.
1 rue de la Fonderie B.P. 1210
F - 68054 Mulhousse

Luis Arimany P.
Idae Campsa EEC
Paseo de la Castellana 95 planta 21
E - Madrid 28046

Luis Ciro Perez
Idae
Paseo de la Castellana 95 planta 21
E - Madrid 28046

Luiz Arnaiz C
Uvisa
Pol.Ind.de Villalonquejar Burgos
Apartado 316
E - Burgos

M & C Clemente
Escuela de Minas
C/Alenza 1
E - Madrid 28003

M. Camino Otero
Enagas
Avda America 38
E - Madrid

M.Cuesta Rubio
Administracion de Servicios
Espana

M.Flores P.
Lonjas y Mercados
Zurbano 74 1 Derecha
E - Madrid

M.Flores S.
Lonjas y Mercados
Zurbano 74 1 Derecha
E - Madrid

M.Jimenez Montes
Foster Wheeler
Espana

M.Ochoa Pelcori
Enagas
Avda America 38
E - Madrid 28028

M.Olive Riu
Catalana de Gas s.a.
Avda Portal de l'Angel 22
E - Barcelone 08002

M.Olive Riu
Catalana de Gas s.a.
Avda Portal de l'Angel 22
E - Barcelone 08002

M.Ruiz Puello
Initec
Alenza 4
E - Madrid

M.Trevena
Energy Technology Support Unit
UK - Abingdon Oxfordshire OX 11 ORA

Mac Glade
Pizier Chemical
Irlande

Manfred Schou
Association of Danish Elect.Utilit.
Rosenorns Allee 9
DK - 1970 Frederiksberg C

Manuel Ruiz P.
Initec
Alenza 4
E - Madris 28003

Mascia Georges
New Projects
F - 72310 Besse sur Braye

Miguel Matey T
Papelera Peninsular s.a.
Paseo de Yeserias 23
E - Madrid 28005

Miguel Matey T
Papelera Peninsular s.a.
Paseo de Yeserias 23
E - Madrid 28005

Molinette A.
Foster Wheeler
Espana

Mortensen H.C.

P.O. Box 304
DK - 2500 Valby

Muratore E;
Centre Technique du Parler
BP 7110
F - 38020 Grenoble

N.Cordeiro d.

rua da Piscina 5 5D
Miraflores Algues
P - Lisboa 1495

Nedergaard N.
Herning Municipal Works
DK - Enghavevej 10

Neumann Gunter

Waldsangerpfad 4
D - 1000 Berlin 38

Nigel G. Foster
Energy Efficiency Office
Thames House Millbank
UK - London SWIP HQ5

P.Garcia Tormo
Sereland
Espana

P.Spindler L.
Elselskabet Effo
Undalsvej 3
DK - 3300 Frederiksvaerk

Pantoja A.
Hidrola
Hermosilis 3
E - 28001 Madrid

Pedro Ledesma
Enagas
Avda America 38
E - Madris 28028

Pierre J.Van Tiggelen
Université Catholique de Louvain
place L.Pasteur 1
B - 1348 Louvain la Neuve

Pierre Jean M.
Gaz de France
23 rue Phillbert Delorme
F - 75017 Paris

Pizzi Roberto

Viale Castello Della Masliana
I - 68 Roma

Poggi Sergio
Ilva Spa
4 via Corcica
I - 16128 Genova

Policarpo Garcia
Carto Espana s.a.
Ctra.Burgos-portugal KM 126
E - Valladolid 47014

Policarpo Garcia
Carto Espana s.a.
Ctra.Burgos - Portugal KM 126
E - Valladolid 47014

Post H. M.
Ests bv Adrescode 2H-14
Kesslerplein 1
NL - 1951 JZ Velsen Noord

Pr. Syred N.
School of Engineering
P.O. Box 917
UK - Cardiff CF2 1XH

R.Arraco M.
Enagas
Avda America 38
E - 28028 Madrid

R.De La Cruz
Hingasa
Espana

R.Diaz Aguero
Forsa
Espana

R.Fer,a,dez L;
Nacional de Gas s.a.
Avda America 38
E - Madrid 28028

R.Fernandez L.
Nacional de Gas s.a. Enagas
Avda America 38
E - Madrid 280

R.Peris Reig
Catalana de Gas s.a.
Avda Portal de l'Angel 22
E - Barcelona 08002

R.Terren B.
Diputacion General de Aragon
P.Maria Augustin 6/N
E - Zaragoza 5

Rahilly G.
Electricity Supply Board
27 LR Fitzwilliam str
IR - Dublin 2 Eire

Richard William Grey

UK - Bracknell Berkshire RG12 4AH

Von Gemeire F.

St Pirherrieumush 41
B - 9000 Gent

Roberto Baviera
Baviera Ahorro de Energia
Gomez Ferrer 33
E - 46900 Vale Torrente

S.Boado Ariza
Vulcano Sadeca
Ctra Vicalvaro Arrises KM 5,6
E - Madrid 28052

S.De La Fuente
Catalana de Gas
Espana

S.Montero
Hingasa
Espana

S.Munoz Gama
Unesa
Espana

S.Vialet
Sereland
Espana

Santiago Feijo
Indein
Guzman el Bueno 133
E - 28003 Madrid

Serrano Fco.
Idae
Paseo de la Castellana 95 pl
E - Madrid 28046

Steimle F.
Universitat Essen
Universitatstrasse 15
D - 4300 Essen 1

T.Fernandes
O.P.E.
P - Lisboa

Tabet J-P.
AFME
27 rue Louis Vicat
F - 75015 Paris

Tatiana Tamayo B
Felguera I.H.I.
Orense 8 Segundo
E - Madrid

Termohlen F.J.
Provinciale Gelderse E.
Holland

Tessitore Elio
T.E. srl
Via Cherubini 15
I - 10154 Torino

Tinbert
Industries Courdes Serete
86 rue Regnault
F - 75640 Paris Cedex 13

Tomas Eric
Université Libre de Bruxelles
av. Roosevelt 50
B - 1050 Bruxelles

Urbano Dominguez
Universidad de Salamanca
E - Bejar 37700

V.Alba Gonzalez
Hidroelectrica Espanola
Espana

Van Hal L.
Soc. des Pétroles Shell
B.P. n°1
F - 76650 Le Petit Couronne

Verbruggen Aviel
University of Antwerp
Prinsstraat 13
B - 2000 Antwerp

INDEX OF AUTHORS

ALBISU, F., 10

BERKELMANS, F.W., 127

CAPARROS, J.J., 112
CONTRERAS, D., 35

DIAZ VARGAS, A., 98
DRISCOLL, D., 85, 98

FEE, A., 74, 98
FOUNAND COLL, C., 181

GOMEZ-ANGULO, A., 35
GREEN, D., 98
GYFTOPOULOS, E.P., 20, 98

HAMRIN, J., 62, 98
HOLLROTTER, E., 151

KINDERMANN, F., 5
KLOP, P.G., 127

LOETH, P., 164

MARANIELLO, G., 117
McGLADE, P.P., 139
MONTALT ROS, L., 181

PEREZ PRIM, J.M., 2

SERRANO, F., 4
SIRCHIS, J., 10, 98

TERMÖHLEN, F.J., 127

For Product Safety Concerns and Information please contact our EU representative GPSR@taylorandfrancis.com
Taylor & Francis Verlag GmbH, Kaufingerstraße 24, 80331 München, Germany

www.ingramcontent.com/pod-product-compliance
Ingram Content Group UK Ltd.
Pitfield, Milton Keynes, MK11 3LW, UK
UKHW011009180926
475701UK00004B/649

9780367580131